the geometry of

W9-BUQ-679

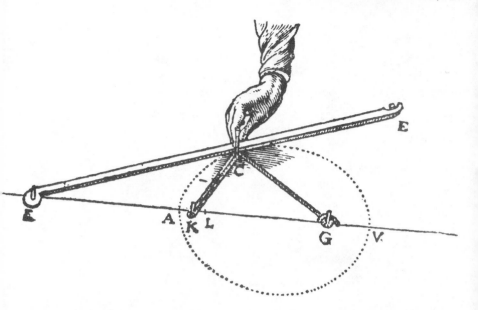

René Descartes

translated from the French and Latin by David Eugene Smith and Marcia L. Latham

Dover Publications, Inc. New York

This Dover edition, first published in 1954, is an unabridged and unaltered republication of the English translation by David Eugene Smith and Marcia L. Latham originally published by the Open Court Publishing Co. in 1925.

International Standard Book Number: 0-486-60068-8

Library of Congress Catalog Card Number: 54-3995

Manufactured in the United States of America
Dover Publications, Inc.
180 Varick Street
New York, N.Y. 10014

Preface

If a mathematician were asked to name the great epoch-making works in his science, he might well hesitate in his decision concerning the product of the nineteenth century; he might even hesitate with respect to the eighteenth century; but as to the product of the sixteenth and seventeenth centuries, and particularly as to the works of the Greeks in classical times, he would probably have very definite views. He would certainly include the works of Euclid, Archimedes, and Apollonius among the products of the Greek civilization, while among those which contributed to the great renaissance of mathematics in the seventeenth century he would as certainly include *La Géométrie* of Descartes and the *Principia* of Newton.

But it is one of the curious facts in the study of historical material that although we have long had the works of Euclid, Archimedes, Apollonius, and Newton in English, the epoch-making treatise of Descartes has never been printed in our language, or, if so, only in some obscure and long-since-forgotten edition. Written originally in French, it was soon after translated into Latin by Van Schooten, and this was long held to be sufficient for any scholars who might care to follow the work of Descartes in the first printed treatise that ever appeared on analytic geometry. At present it is doubtful if many mathematicians read the work in Latin; indeed, it is doubtful if many except the French scholars consult it very often in the original language in which it appeared. But certainly a work of this kind ought to be easily accessible to American and British students of the history of mathematics, and in a language with which they are entirely familiar.

On this account, The Open Court Publishing Company has agreed with the translators that the work should appear in English, and with such notes as may add to the ease with which it will be read. To this organization the translators are indebted for the publication of the book, a labor of love on its part as well as on theirs.

As to the translation itself, an attempt has been made to give the meaning of the original in simple English rather than to add to the difficulty of the reader by making it a verbatim reproduction. It is believed that the student will welcome this policy, being content to go to the original in case a stricter translation is needed. One of the translators having used chiefly the Latin edition of Van Schooten, and the other the original French edition, it is believed that the meaning which Descartes had in mind has been adequately preserved.

Table of Contents[1]

BOOK I

PROBLEMS THE CONSTRUCTION OF WHICH REQUIRES ONLY STRAIGHT
LINES AND CIRCLES

---

[1] It should be recalled that the first edition of this work appeared as a kind of appendix to the *Discours de la Methode,* and hence began on page 297. For convenience of reference, the original paging has been retained in the facsimile. A new folio number, appropriate to the present edition, will also be found at the foot of each page. For convenience of reference to the original, this table of contents follows the paging of the 1637 edition.

TABLE

Des matieres de la

GEOMETRIE.

Liure Premier.

DES PROBLESMES QU'ON PEUT
conftruire fans y employer que des cercles &
des lignes droites.

Kkk   Com

BOOK II

On the Nature of Curved Lines

Discours Second.
DE LA NATURE DES LIGNES
COURBES.

Com~

BOOK III

ON THE CONSTRUCTION OF SOLID OR SUPERSOLID PROBLEMS

Liure Troisiesme
DE LA CONSTRUCTION DES
problesmes solides, ou plusque solides.

TABLE OF CONTENTS

F I N.

Les

BOOK FIRST

The Geometry of René Descartes

BOOK I

PROBLEMS THE CONSTRUCTION OF WHICH REQUIRES ONLY STRAIGHT LINES AND CIRCLES

ANY problem in geometry can easily be reduced to such terms that a knowledge of the lengths of certain straight lines is sufficient for its construction.[1] Just as arithmetic consists of only four or five operations, namely, addition, subtraction, multiplication, division and the extraction of roots, which may be considered a kind of division, so in geometry, to find required lines it is merely necessary to add or subtract other lines; or else, taking one line which I shall call unity in order to relate it as closely as possible to numbers,[2] and which can in general be chosen arbitrarily, and having given two other lines, to find a fourth line which shall be to one of the given lines as the other is to unity (which is the same as multiplication); or, again, to find a fourth line which is to one of the given lines as unity is to the other (which is equivalent to division); or, finally, to find one, two, or several mean proportionals between unity and some other line (which is the same

---

[1] Large collections of problems of this nature are contained in the following works: Vincenzo Riccati and Girolamo Saladino, *Institutiones Analyticae,* Bologna, 1765; Maria Gaetana Agnesi, *Istituzioni Analitiche,* Milan, 1748; Claude Rabuel, *Commentaires sur la Géométrie de M. Descartes,* Lyons, 1730 (hereafter referred to as Rabuel); and other books of the same period or earlier.

[2] Van Schooten, in his Latin edition of 1683, has this note: "Per unitatem intellige lineam quandam determinatam, qua ad quamvis reliquarum linearum talem relationem habeat, qualem unitas ad certum aliquem numerum." *Geometria a Renato Des Cartes, una cum notis Florimondi de Beaune, opera atque studio Francisci à Schooten,* Amsterdam, 1683, p. 165 (hereafter referred to as Van Schooten).

In general, the translation runs page for page with the facing original. On account of figures and footnotes, however, this plan is occasionally varied, but not in such a way as to cause the reader any serious inconvenience.

LA
GEOMETRIE.
LIVRE PREMIER.

Des problefmes qu'on peut conftruire fans
y employer que des cercles & des
lignes droites.

TOus les Problefmes de Geometrie fe
peuuent facilement reduire a tels termes,
qu'il n'eft befoin par aprés que de connoiftre la longeur de quelques lignes droites,
pour les conftruire.

Et comme toute l'Arithmetique n'eft compofée, que
de quatre ou cinq operations, qui font l'Addition, la
Souftraction, la Multiplication, la Diuifion, & l'Extraction des racines, qu'on peut prendre pour vne efpece
de Diuifion : Ainfi n'at'on autre chofe a faire en Geometrie touchant les lignes qu'on cherche, pour les pre-
parer a eftre connuës, que leur en adioufter d'autres, ou
en ofter, Oubien en ayant vne, que ie nommeray l'vnité
pour la rapporter d'autant mieux aux nombres, & qui
peut ordinairement eftre prife a difcretion, puis en ayant
encore deux autres, en trouuer vne quatriefme, qui foit
à l'vne de ces deux, comme l'autre eft a l'vnité, ce qui eft
le mefme que la Multiplication ; oubien en trouuer vne
quatriefme, qui foit a l'vne de ces deux, comme l'vnité

Commēt
le calcul
d'Ari-
thmeti-
que fe
rapporte
aux ope-
rations de
Geome-
trie.

eft

eſt a l'autre, ce qui eſt le meſme que la Diuiſion; ou enfin trouuer vne, ou deux, ou pluſieurs moyennes proportionnelles entre l'vnité, & quelque autre ligne; ce qui eſt le meſme que tirer la racine quarrée, on cubique,&c. Et ie ne craindray pas d'introduire ces termes d'Arithmetique en la Geometrie, affin de me rendre plus intelligibile.

La Multiplication.

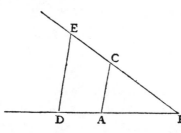

Soit par exemple A B l'vnité, & qu'il faille multiplier B D par B C, ie n'ay qu'a ioindre les poins A & C, puis tirer D E parallele a C A, & B E eſt le produit de cete Multiplication.

La Diviſion.

Oubien s'il faut diuiſer B E par B D, ayant ioint les poins E & D, ie tire A C parallele a D E, & B C eſt le produit de cete diuiſion.

l'Extraction de la racine quarrée.

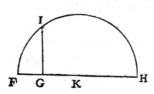

Ou s'il faut tirer la racine quarrée de G H, ie luy adiouſte en ligne droite F G, qui eſt l'vnité, & diuiſant F H en deux parties eſgales au point K, du centre K ie tire le cercle F I H, puis eſleuant du point G vne ligne droite iuſques à I, à angles droits ſur F H, c'eſt G I la racine cherchée. Ie ne dis rien icy de la racine cubique, ny des autres, à cauſe que i'en parleray plus commodement cy aprés.

Commēt on peut

Mais ſouuent on n'a pas beſoin de tracer ainſi ces ligne

as extracting the square root, cube root, etc., of the given line.[3] And I shall not hesitate to introduce these arithmetical terms into geometry, for the sake of greater clearness.

For example, let AB be taken as unity, and let it be required to multiply BD by BC. I have only to join the points A and C, and draw DE parallel to CA; then BE is the product of BD and BC.

If it be required to divide BE by BD, I join E and D, and draw AC parallel to DE; then BC is the result of the division.

If the square root of GH is desired, I add, along the same straight line, FG equal to unity; then, bisecting FH at K, I describe the circle FIH about K as a center, and draw from G a perpendicular and extend it to I, and GI is the required root. I do not speak here of cube root, or other roots, since I shall speak more conveniently of them later.

Often it is not necessary thus to draw the lines on paper, but it is sufficient to designate each by a single letter. Thus, to add the lines BD and GH, I call one a and the other b, and write $a + b$. Then $a - b$ will indicate that b is subtracted from a; ab that a is multiplied by b; $\frac{a}{b}$ that a is divided by b; aa or a^2 that a is multiplied by itself; a^3 that this result is multiplied by a, and so on, indefinitely.[4] Again, if I wish to extract the square root of a^2+b^2, I write $\sqrt{a^2+b^2}$; if I wish to extract the cube root of $a^3-b^3+ab^2$, I write $\sqrt[3]{a^3-b^3+ab^2}$, and similarly for other roots.[5] Here it must be observed that by a^2, b^3, and similar expressions, I ordinarily mean only simple lines, which, however, I name squares, cubes, etc., so that I may make use of the terms employed in algebra.[6]

[3] While in arithmetic the only exact roots obtainable are those of perfect powers, in geometry a length can be found which will represent exactly the square root of a given line, even though this line be not commensurable with unity. Of other roots, Descartes speaks later.

[4] Descartes uses a^3, a^4, a^5, a^6, and so on, to represent the respective powers of a, but he uses both aa and a^2 without distinction. For example, he often has $aabb$, but he also uses $\frac{3a^2}{4b^2}$.

[5] Descartes writes: $\sqrt{C.a^3 - b^3 + abb}$. See original, page 299, line 9.

[6] At the time this was written, a^2 was commonly considered to mean the surface of a square whose side is a, and b^3 to mean the volume of a cube whose side is b; while b^4, b^5, ... were unintelligible as geometric forms. Descartes here says that a^2 does not have this meaning, but means the line obtained by constructing a third proportional to 1 and a, and so on.

It should also be noted that all parts of a single line should always be expressed by the same number of dimensions, provided unity is not determined by the conditions of the problem. Thus, a^3 contains as many dimensions as ab^2 or b^3, these being the component parts of the line which I have called $\sqrt[3]{a^3-b^3+ab^2}$. It is not, however, the same thing when unity is determined, because unity can always be understood, even where there are too many or too few dimensions; thus, if it be required to extract the cube root of $a^2b^2 - b$, we must consider the quantity a^2b^2 divided once by unity, and the quantity b multiplied twice by unity.[7]

Finally, so that we may be sure to remember the names of these lines, a separate list should always be made as often as names are assigned or changed. For example, we may write, AB=1, that is AB is equal to 1 ;[8] GH $= a$, BD $= b$, and so on.

If, then, we wish to solve any problem, we first suppose the solution already effected,[9] and give names to all the lines that seem needful for its construction,—to those that are unknown as well as to those that are known.[10] Then, making no distinction between known and unknown lines, we must unravel the difficulty in any way that shows most natur-

[7] Descartes seems to say that each term must be of the third degree, and that therefore we must conceive of both a^2b^2 and b as reduced to the proper dimension.

[8] Van Schooten adds "seu unitati," p. 3. Descartes writes, AB ∞ 1. He seems to have been the first to use this symbol. Among the few writers who followed him, was Hudde (1633-1704). It is very commonly supposed that ∞ is a ligature representing the first two letters (or diphthong) of "æquare." See, for example, M. Aubry's note in W. W. R. Ball's *Recréations Mathématiques et Problèmes des Temps Anciens et Modernes*, French edition, Paris, 1909, Part III, p. 164.

[9] This plan, as is well known, goes back to Plato. It appears in the work of Pappus as follows: "In analysis we suppose that which is required to be already obtained, and consider its connections and antecedents, going back until we reach either something already known (given in the hypothesis), or else some fundamental principle (axiom or postulate) of mathematics." *Pappi Alexandrini Collectiones quae supersunt e libris manu scriptis edidit Latina interpellatione et commentariis instruxit Fredericus Hultsch*, Berlin, 1876-1878; vol. II, p. 635 (hereafter referred to as Pappus). See also Commandinus, *Pappi Alexandrini Mathematicae Collectiones*, Bologna, 1588, with later editions.

Pappus of Alexandria was a Greek mathematician who lived about 300 A.D. His most important work is a mathematical treatise in eight books, of which the first and part of the second are lost. This was made known to modern scholars by Commandinus. The work exerted a happy influence on the revival of geometry in the seventeenth century. Pappus was not himself a mathematician of the first rank, but he preserved for the world many extracts or analyses of lost works, and by his commentaries added to their interest.

[10] Rabuel calls attention to the use of a, b, c, ... for known, and x, y, z, ... for unknown quantities (p. 20).

gnes fur le papier, & il fuffift de les defigner par quelques vfer de chiffres en Geome- trie. lettres, chafcune par vne feule. Comme pour adioufter la ligne B D a G H, ie nomme l'vne *a* & l'autre *b*, & efcris *a* +- *b*; Et *a* -- *b*, pour fouftraire *b* d' *a*; Et *a b*, pour les multiplier l'vne par l'autre; Et $\frac{a}{b}$, pour diuifer *a* par *b* ; Et *a a*,

ou $\overset{2}{a}$, pour multiplier *a* par foy mefme ; Et $\overset{3}{a}$, pour le multiplier encore vne fois par *a* , & ainfi a l'infini ; Et

$\sqrt{\overset{2}{a} +- \overset{2}{b}}$, pour tirer la racine quarrée d' $\overset{2}{a}$ +- $\overset{2}{b}$; Et

$\sqrt{C. \overset{3}{a} -- \overset{3}{b} +- a b b}$, pour tirer la racine cubique d' $\overset{3}{a}$ -- $\overset{3}{b}$ +- *a b b*, & ainfi des autres.

Où il eft a remarquer que par $\overset{2}{a}$ ou $\overset{3}{b}$ ou femblables, ie ne conçoy ordinairement que des lignes toutes fimples, encore que pour me feruir des noms vfités en l'Algebre, ie les nomme des quarrés ou des cubes, &c.

Il eft auffy a remarquer que toutes les parties d'vne mefme ligne, fe doiuent ordinairement exprimer par autant de dimenfions l'vne que l'autre, lorfque l'vnité n'eft point déterminée en la queftion, comme icy $\overset{3}{a}$ en contient autant qu' *a b b* ou $\overset{3}{b}$ dont fe compofe la ligne que i'ay nommée $\sqrt{C. \overset{3}{a} -- \overset{3}{b} +- a b b}$: mais que ce n'eft pas de mefme lorfque l'vnité eft déterminée, a caufe qu'elle peut eftre foufentendue par tout ou il y a trop ou trop peu de dimenfions : comme s'il faut tirer la racine cubique de *a a b b* -- *b* , il faut penfer que la quantité *a a b b* eft diuifée vne fois par l'vnité, & que l'autre quantité *b* eft multipliée deux fois par la mefme.

Au refte affin de ne pas manquer a fe fouuenir des noms de ces lignes, il en faut toufiours faire vn regiftre feparé , à mefure qu'on les pofe ou qu'on les change, efcriuant par exemple.

A B ∞ 1,   c'eft a dire, A B efgal à 1.

G H ∞ *a*

BD ∞ *b*, &c.

Commēt il faut ve- nir aux Equatiõs qui fer- uent a re- foudre les problef- mes. Ainfi voulant refoudre quelque problefme, on doit d'a- bord le confiderer comme defia fait, & donner des noms a toutes les lignes, qui femblent neceffaires pour le con- ftruire, auffy bien a celles qui font inconnuës, qu'aux autres. Puis fans confiderer aucune difference entre ces lignes connuës, & inconnuës, on doit parcourir la diffi- culté, felon l'ordre qui monftre le plus naturellement de tous en qu'elle forte elles dependent mutuellement. les vnes des autres, iufques a ce qu'on ait trouué moyen `d'exprimer vne mefme quantité en deux façons: ce qui fe nomme vne Equation; car les termes de l'vne de ces deux façons font efgaux a ceux de l'autre.   Et on doit trouuer autant de telles Equations, qu'on a fuppofé de li- gnes, qui eftoient inconnuës.   Oubien s'il ne s'en trouue pas tant, & que nonobftant on n'omette rien de ce qui eft defiré en la queftion, cela tefmoigne qu'elle n'eft pas en- tierement determinée. Et lors on peut prendre a difcre- tion des lignes connuës, pour toutes les inconnuës auf- qu'elles ne correfpond aucune Equation. Aprés cela s'il en refte encore plufieurs, il fe faut feruir par ordre de chafcune des Equations qui reftent auffy, foit en la con- fiderant toute feule, foit en la comparant auec les autres, pour expliquer chafcune de ces lignes inconnuës; & faire
ainfi

ally the relations between these lines, until we find it possible to express a single quantity in two ways.[11] This will constitute an equation, since the terms of one of these two expressions are together equal to the terms of the other.

We must find as many such equations as there are supposed to be unknown lines;[12] but if, after considering everything involved, so many cannot be found, it is evident that the question is not entirely determined. In such a case we may choose arbitrarily lines of known length for each unknown line to which there corresponds no equation.[13]

If there are several equations, we must use each in order, either considering it alone or comparing it with the others, so as to obtain a value for each of the unknown lines; and so we must combine them until there remains a single unknown line[14] which is equal to some known line, or whose square, cube, fourth power, fifth power, sixth power, etc., is equal to the sum or difference of two or more quantities,[15] one of which is known, while the others consist of mean proportionals between unity and this square, or cube, or fourth power, etc., multiplied by other known lines. I may express this as follows:

$$z = b,$$
$$\text{or } z^2 = -az + b^2,$$
$$\text{or } z^3 = az^2 + b^2z - c^3,$$
$$\text{or } z^4 = az^3 - c^3z + d^4, \text{ etc.}$$

That is, z, which I take for the unknown quantity, is equal to b; or, the square of z is equal to the square of b diminished by a multiplied by z; or, the cube of z is equal to a multiplied by the square of z, plus the square of b multiplied by z, diminished by the cube of c; and similarly for the others.

[11] That is, we must solve the resulting simultaneous equations.

[12] Van Schooten (p. 149) gives two problems to illustrate this statement. Of these, the first is as follows: Given a line segment AB containing any point C, required to produce AB to D so that the rectangle AD.DB shall be equal to the square on CD. He lets $AC = a$, $CB = b$, and $BD = x$. Then $AD = a + b + x$, and $CD = b + x$, whence $ax + bx + x^2 = b^2 + 2bx + x^2$ and $x = \dfrac{b^2}{a-b}$.

[13] Rabuel adds this note: "We may say that every indeterminate problem is an infinity of determinate problems, or that every problem is determined either by itself or by him who constructs it" (p. 21).

[14] That is, a line represented by x, x^2, x^3, x^4, ...

[15] In the older French, "le quarré, ou le cube, ou le quarré de quarré, ou le sursolide, ou le quarré de cube &c.," as seen on page 11 (original page 302).

Thus, all the unknown quantities can be expressed in terms of a single quantity,[16] whenever the problem can be constructed by means of circles and straight lines, or by conic sections, or even by some other curve of degree not greater than the third or fourth.[17]

But I shall not stop to explain this in more detail, because I should deprive you of the pleasure of mastering it yourself, as well as of the advantage of training your mind by working over it, which is in my opinion the principal benefit to be derived from this science. Because, I find nothing here so difficult that it cannot be worked out by any one at all familiar with ordinary geometry and with algebra, who will consider carefully all that is set forth in this treatise.[18]

[16] See line 20 on the opposite page.

[17] Literally, "Only one or two degrees greater."

[18] In the Introduction to the 1637 edition of *La Géométrie*, Descartes made the following remark: "In my previous writings I have tried to make my meaning clear to everybody; but I doubt if this treatise will be read by anyone not familiar with the books on geometry, and so I have thought it superfluous to repeat demonstrations contained in them." See *Oeuvres de Descartes*, edited by Charles Adam and Paul Tannery, Paris, 1897-1910, vol. VI, p. 368. In a letter written to Mersenne in 1637 Descartes says: "I do not enjoy speaking in praise of myself, but since few people can understand my geometry, and since you wish me to give you my opinion of it, I think it well to say that it is all I could hope for, and that in *La Dioptrique* and *Les Météores*, I have only tried to persuade people that my method is better than the ordinary one. I have proved this in my geometry, for in the beginning I have solved a question which, according to Pappus, could not be solved by any of the ancient geometers.

"Moreover, what I have given in the second book on the nature and properties of curved lines, and the method of examining them, is, it seems to me, as far beyond the treatment in the ordinary geometry, as the rhetoric of Cicero is beyond the a, b, c of children. . . .

"As to the suggestion that what I have written could easily have been gotten from Vieta, the very fact that my treatise is hard to understand is due to my attempt to put nothing in it that I believed to be known either by him or by any one else. . . . I begin the rules of my algebra with what Vieta wrote at the very end of his book, *De emendatione aequationum*. . . . Thus, I begin where he left off." *Oeuvres de Descartes, publiées par Victor Cousin*, Paris, 1824, Vol. VI, p. 294 (hereafter referred to as Cousin).

In another letter to Mersenne, written April 20, 1646, Descartes writes as follows: "I have omitted a number of things that might have made it (the geometry) clearer, but I did this intentionally, and would not have it otherwise. The only suggestions that have been made concerning changes in it are in regard to rendering it clearer to readers, but most of these are so malicious that I am completely disgusted with them." Cousin, Vol. IX, p. 553.

In a letter to the Princess Elizabeth, Descartes says: "In the solution of a geometrical problem I take care, as far as possible, to use as lines of reference parallel lines or lines at right angles; and I use no theorems except those which assert that the sides of similar triangles are proportional, and that in a right triangle the square of the hypotenuse is equal to the sum of the squares of the sides. I do not hesitate to introduce several unknown quantities, so as to reduce the question to such terms that it shall depend only on these two theorems." Cousin, Vol. IX, p. 143.

10

ainfi en les demeflant, qu'il n'en demeure qu'vne feule,
efgale a quelque autre, qui foit connuë, oubien dont le
quarré, ou le cube, ou le quarré de quarré, ou le furfolide, ou le quarré de cube, &c. foit efgal a ce, qui fe pro-
duift par l'addition, ou fouftraction de deux ou plufieurs
autres quantités, dont l'vne foit connuë, & les autres
foient compofées de quelques moyennes proportionnelles entre l'vnité, & ce quarré, ou cube, ou quarré de
quarré, &c. multipliées par d'autres connuës. Ce que i'e-
fcris en cete forte.

$$z \infty b. \text{ ou}$$

$$z^2 \infty -- a\, z + bb. \text{ ou}$$

$$z^3 \infty + a\, z^2 + bb\, z -- c. \text{ ou}$$

$$z^4 \infty a\, z^3 - c\, z + d^4. \&c.$$

C'eft a dire, z, que ie prens pour la quantité inconnuë,
eft efgalé a b, ou le quarré de z eft efgal au quarré de b
moins a multiplié par z. ou le cube de z eft efgal à a
multiplié par le quarre de z plus le quarré de b multiplié
par z moins le cube de c. & ainfi des autres.

Et on peut toufiours reduire ainfi toutes les quantités
inconnuës à vne feule, lorfque le Problefme fe peut conftruire par des cercles & des lignes droites, ou auffy par
des fections coniques, ou mefme par quelque autre ligne
qui ne foit que d'vn ou deux degrés plus compofée. Mais
ie ne m'arefte point a expliquer cecy plus en detail, a
caufe que ie vous ofterois le plaifir de l'apprendre de
vous mefme, & l'vtilité de cultiuer voftre efprit en vous
y exerceant, qui eft a mon auis la principale, qu'on puiffe

Pp 3   tirer

tırer de cete fcience. Auffy que ie n y remarque rien de
fi difficile, que ceux qui feront vn peu verfés en la Geometrie commune, & en l'Algebre, & qui prendront garde a tout ce qui eſt en ce traité, ne puiſſent trouuer.

C'eſt pourquoy ie me contenteray icy de vous auertir, que pourvû qu'en demeſlant ces Equations on ne
manque point a ſe ſeruir de toutes les diuiſions, qui ſeront poſſibles, on aura infalliblement les plus ſimples
termes, auſquels la queſtion puiſſe eſtre reduite.

Quels
ſont les
problef-
mes plans Et que ſi elle peut eſtre reſolue par la Geometrie ordinaire, c'eſt a dire, en ne ſe ſeruant que de lignes droites
& circulaires tracées ſur vne ſuperficie plate, lorſque la
derniere Equation aura eſté entierement démeſlée, il n'y
reſtera tout au plus qu'vn quarré inconnu, eſgal a ce qui
ſe produiſt de l'Addition, ou ſouſtraction de ſa rácine
multipliée par quelque quantité connue, & de quelque
autre quantité auſſy connue

Com-
ment ils
ſe reſol-
uent. Et lors cete racine, ou ligne inconnue ſe trouue ayſement. Car ſi i'ay par exemple

$$z \infty\, a\, z + b\, b$$

ie fais le triangle rectangle N L M, dont le co-
ſté L M eſt eſgal à *b* racine quarrée de la quan-
tité connue *b b*, & l'autre L N eſt ½ *a*, la moitié de l'autre quantité
connue, qui eſtoit multipliée par *z* que ie ſuppoſe eſtre la
ligne inconnue. puis prolongeant M N la baze de ce triangle,

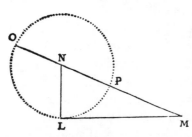

I shall therefore content myself with the statement that if the student, in solving these equations, does not fail to make use of division wherever possible, he will surely reach the simplest terms to which the problem can be reduced.

And if it can be solved by ordinary geometry, that is, by the use of straight lines and circles traced on a plane surface,[19] when the last equation shall have been entirely solved there will remain at most only the square of an unknown quantity, equal to the product of its root by some known quantity, increased or diminished by some other quantity also known.[20] Then this root or unknown line can easily be found. For example, if I have $z^2 = az + b^2$,[21] I construct a right triangle NLM with one side LM, equal to b, the square root of the known quantity b^2, and the other side, LN, equal to $\frac{1}{2}a$, that is, to half the other known quantity which was multiplied by z, which I supposed to be the unknown line. Then prolonging MN, the hypotenuse[22] of this triangle, to O, so that NO is equal to NL, the whole line OM is the required line z. This is expressed in the following way:[23]

$$z = \frac{1}{2}a + \sqrt{\frac{1}{4}a^2 + b^2}.$$

But if I have $y^2 = -ay + b^2$, where y is the quantity whose value is desired, I construct the same right triangle NLM, and on the hypote-

[19] For a discussion of the possibility of constructions by the compasses and straight edge, see Jacob Steiner, *Die geometrischen Constructionen ausgeführt mittelst der geraden Linie und eines festen Kreises*, Berlin, 1833. For briefer treatments, consult Enriques, *Fragen der Elementar-Geometrie*, Leipzig, 1907; Klein, *Problems in Elementary Geometry*, trans. by Beman and Smith, Boston, 1897; Weber und Wellstein, *Encyklopädie der Elementaren Geometrie*, Leipzig, 1907. The work by Mascheroni, *La geometria del compasso*, Pavia, 1797, is interesting and well known.

[20] That is, an expression of the form $z^2 = az \pm b$. "Esgal a ce qui se produit de l'Addition, ou soustraction de sa racine multiplée par quelque quantité connue, & de quelque autre quantité aussy connue," as it appears in line 14, opposite page.

[21] Descartes proposes to show how a quadratic may be solved geometrically.

[22] Descartes says "prolongeant MN la baze de ce triangle," because the hypotenuse was commonly taken as the base in earlier times.

[23] From the figure $OM \cdot PM = LM^2$. If $OM = z$, $PM = z - a$, and since $LM = b$, we have $z(z-a) = b^2$ or $z^2 = az + b^2$. Again, $MN = \sqrt{\frac{1}{4}a^2 + b^2}$, whence $OM = z = ON + MN = \frac{1}{2}a + \sqrt{\frac{1}{4}a^2 + b^2}$. Descartes ignores the second root, which is negative.

nuse MN lay off NP equal to NL, and the remainder PM is y, the desired root. Thus I have

$$y = -\frac{1}{2}a + \sqrt{\frac{1}{4}a^2 + b^2}.$$

In the same way, if I had

$$x^4 = -ax^2 + b^2,$$

PM would be x^2 and I should have

$$x = \sqrt{-\frac{1}{2}a + \sqrt{\frac{1}{4}a^2 + b^2}},$$

and so for other cases.

Finally, if I have $z^2 = az - b^2$, I make NL equal to $\frac{1}{2}a$ and LM equal to b as before; then, instead of joining the points M and N, I draw MQR parallel to LN, and with N as a center describe a circle through L cutting MQR in the points Q and R; then z, the line sought, is either MQ or MR, for in this case it can be expressed in two ways, namely:[24]

$$z = \frac{1}{2}a + \sqrt{\frac{1}{4}a^2 - b^2},$$

and

$$z = \frac{1}{2}a - \sqrt{\frac{1}{4}a^2 - b^2}.$$

---

[24] Since $MR.MQ = \overline{LM}^2$, then if $R = z$, we have $MQ = a - z$, and so $z(a - z) = b^2$ or $z^2 = az - b^2$.

If, instead of this, $MQ = z$, then $MR = a - z$, and again, $z^2 = az - b^2$. Furthermore, letting O be the mid-point of QR,

$$MQ = OM - OQ = \frac{1}{2}a - \sqrt{\frac{1}{4}a^2 - b^2},$$

and

$$MR = MO + OR = \frac{1}{2}a + \sqrt{\frac{1}{4}a^2 - b^2}.$$

Descartes here gives both roots, since both are positive. If MR is tangent to the circle, that is, if $b = \frac{1}{2}a$, the roots will be equal; while if $b > \frac{1}{2}a$, the line MR will not meet the circle and both roots will be imaginary. Also, since $RM.QM = \overline{LM}^2$, $z_1 z_2 = b^2$, and $RM + QM = z_1 + z_2 = a$.

angle, iufques a O, en forte qu'N O foit efgale a N L, la toute O M eft z la ligne cherchée. Et elle s'exprime en cete forte

$$z \propto \tfrac{1}{2} a + \sqrt{\tfrac{1}{4} a a + b b}.$$

Que fi iay $yy \propto -- a y + b b$, & qu'y foit la quantité qu'il faut trouuer , ie fais le mefme triangle rectangle N L M, & de fa baze M N i'ofte N P efgale a N L, & le refte P M eft y la racine cherchée. De façon que iay

$$y \propto -- \tfrac{1}{2} a + \sqrt{\tfrac{1}{4} a a + b b}.$$ Et tout de mefme fi i'a-

uois $\overset{4}{x} \propto -- a \overset{2}{x} + \overset{2}{b}.$ P M feroit $\overset{2}{x}$. & i'aurois

$$x \propto \sqrt{-- \tfrac{1}{2} a + \sqrt{\tfrac{1}{4} a a + b b}} :$$ & ainfi des autres.

Enfin fi i'ay

$$\overset{2}{z} \propto a z -- b b :$$

ie fais N L efgale à $\tfrac{1}{2} a$, & L M

efgale à b cóme deuãt, puis, au lieu de ioindre les poins M N , ie tire M Q R parallele a L N. & du centre N par L ayant defcrit vn cercle qui la couppe aux poins Q & R, la ligne cherchée z eft M Q, oubiẽ M R, car en ce cas elle s'exprime en deux façons, a fçauoir $z \propto \tfrac{1}{2} a + \sqrt{\tfrac{1}{4} a a -- b b}$, & $z \propto \tfrac{1}{2} a -- \sqrt{\tfrac{1}{4} a a -- b b}.$

Et fi le cercle, qui ayant fon centre au point N , paffe par le point L, ne couppe ny ne touche la ligne droite M Q R, il n'y a aucune racine en l'Equation, de façon qu'on peut affurer que la conftruction du problefme propofé eft impoffible.

Au

Au refte ces mefmes racines fe peuuent trouuer par
vne infinité d'autres moyens , & i'ay feulement veulu
mettre ceux cy, comme fort fimples, affin de faire voir
qu'on peut conftruire tous les Problefmes de la Geometrie ordinaire, fans faire autre chofe que le peu qui eft
compris dans les quatre figures que i'ay expliquées.　Ce
que ie ne croy pas que les anciens ayent remarqué, car
autrement ils n'euffent pas pris la peine d'en efcrire tant
de gros liures, ou le feul ordre de leurs propofitions nous
fait connoiftre qu'ils n'ont point eu la vraye methode
pour les trouuer toutes, mais qu'ils ont feulement ramaf-
fé celles qu'ils ont rencontrées.

Exemple tiré de Pappus.　Et on le peut voir auffy fort clairement de ce que Pappus a mis au commencement de fon feptiefme liure, ou
aprés s'eftre arefté quelque tems a denombrer tout ce
qui auoit efté efcrit en Geometrie par ceux qui l'auoient
precedé, il parle enfin d vne queftion , qu'il dit que ny
Euclide, ny Apollonius, ny aucun autre n'auoient fceu
entierement refoudre. & voycy fes mots.

Ie cite plutoft la verfion la- tine que le texte grec affin que chafcun l'entende plus ayfe- ment.　*Quem autem dicit (Apollonius) in tertio libro locum ad
tres, & quatuor lineas ab Euclide perfeɛtum non effe , neque
ipfe perficere poterat , neque aliquis alius᾿: fed neque paululum quid addere iis , quæ Euclides fcripfit, per ea tantum
conica , quæ ufque ad Euclidis tempora præmonftrata
funt, &c.*

Et vn peu aprés il explique ainfi qu'elle eft cete queftion.

*At locus ad tres, & quatuor lineas, in quo (Apollonius)
magnifice fe iaɛtat, & oftentat, nulla habita gratia ei , qui
prius fcripferat , eft hujufmodi. Si pofitione datis tribus
reɛtis*

16

And if the circle described about N and passing through L neither cuts nor touches the line MQR, the equation has no root, so that we may say that the construction of the problem is impossible.

These same roots can be found by many other methods,[25] I have given these very simple ones to show that it is possible to construct all the problems of ordinary geometry by doing no more than the little covered in the four figures that I have explained.[26] This is one thing which I believe the ancient mathematicians did not observe, for otherwise they would not have put so much labor into writing so many books in which the very sequence of the propositions shows that they did not have a sure method of finding all,[27] but rather gathered together those propositions on which they had happened by accident.

This is also evident from what Pappus has done in the beginning of his seventh book,[28] where, after devoting considerable space to an enumeration of the books on geometry written by his predecessors,[29] he finally refers to a question which he says that neither Euclid nor Apollonius nor any one else had been able to solve completely;[30] and these are his words:

"Quem autem dicit (Apollonius) in tertio libro locum ad tres, & quatuor lineas ab Euclide perfectum non esse, neque ipse perficere poterat, neque aliquis alius; sed neque paululum quid addere iis, quæ

[25] For interesting contraction, see Rabuel, p. 23, et seq.

[26] It will be seen that Descartes considers only three types of the quadratic equation in z, namely, $z^2 + az - b^2 = 0$, $z^2 - az - b^2 = 0$, and $z^2 - az + b^2 = 0$. It thus appears that he has not been able to free himself from the old traditions to the extent of generalizing the meaning of the coefficients, — as negative and fractional as well as positive. He does not consider the type $z^2 + az + b^2 = 0$, because it has no positive roots.

[27] "Qu'ils n'ont point eu la vraye methode pour les trouuer toutes."

[28] See Note [9].

[29] See Pappus, Vol. II, p. 637. Pappus here gives a list of books that treat of analysis, in the following words: "Illorum librorum, quibus de loco, 'αναλυόμενος sive resoluto agitur, ordo hic est. Euclidis datorum liber unus, Apollonii de proportionis sectione libri duo, de spatii sectione duo, de sectione determinata duo, de tactionibus duo, Euclidis porismatum libri tres, Apollonii inclinationum libri duo, eiusdem locorum planorum duo, conicorum octo, Aristaei locorum solidorum libri duo." See also the Commandinus edition of Pappus, 1660 edition, pp. 240-252.

[30] For the history of this problem, see Zeuthen: *Die Lehre von den Kegelschnitten im Alterthum*, Copenhagen, 1886. Also, Adam and Tannery, *Oeuvres de Descartes*, vol. 6, p. 723.

Euclides scripsit, per ea tantum conica, quæ usque ad Euclidis tempora præmonstrata sunt, &c." [31]

A little farther on, he states the question as follows:

"At locus ad tres, & quatuor lineas, in quo (Apollonius) magnifice se jactat, & ostentat, nulla habita gratia ei, qui prius scripserat, est hujusmodi.[32] *Si positione datis tribus rectis lineis ab uno & eodem puncto, ad tres lineas in datis angulis rectæ lineæ ducantur, & data sit proportio rectanguli contenti duabus ductis ad quadratum reliquæ: punctum contingit positione datum solidum locum, hoc est unam ex tribus conicis sectionibus. Et si ad quatuor rectas lineas positione datas in datis angulis lineæ ducantur; & rectanguli duabus ductis contenti ad contentum duabus reliquis proportio data sit; similiter punctum datum coni sectionem positione continget. Si quidem igitur ad duas tantum locus planus ostensus est. Quod si ad plures quam quatuor, punctum continget locos non adhuc cognitos, sed lineas tantum dictas; quales autem sint, vel quam habeant proprietatem, non constat: earum unam, neque primam, & quæ manifestissima videtur, composuerunt ostendentes utilem esse. Propositiones autem ipsarum hæ sunt.*

"Si ab aliquo puncto ad positione datas rectas lineas quinque ducantur rectæ lineæ in datis angulis, & data sit proportio solidi parallelepipedi rectanguli, quod tribus ductis lineis continetur ad solidum parallelepipedum rectangulum, quod continetur reliquis duabus, & data quapiam linea, punctum positione datam lineam continget. Si autem ad sex, & data sit proportio solidi tribus lineis contenti ad solidum, quod tribus reliquis continetur; rursus punctum continget positione datam lineam. Quod si ad plures quam sex, non adhuc habent dicere, an data sit proportio cujuspiam contenti quatuor lineis ad id quod reliquis continetur,

[31] Pappus, Vol. II, pp. 677, et seq., Commandinus edition of 1660, p. 251. Literally, "Moreover, he (Apollonius) says that the problem of the locus related to three or four lines was not entirely solved by Euclid, and that neither he himself, nor any one else has been able to solve it completely, nor were they able to add anything at all to those things which Euclid had written, by means of the conic sections only which had been demonstrated before Euclid." Descartes arrived at the solution of this problem four years before the publication of his geometry, after spending five or six weeks on it. See his letters, Cousin, Vol. VI, p. 294, and Vol. VI, p. 224.

[32] Given as follows in the edition of Pappus by Hultsch, previously quoted: "Sed hic ad tres et quatuor lineas locus quo magnopere gloriatur simul addens ei qui conscripserit gratiam habendam esse, sic se habet."

rectis lineis ab uno & eodem puncto, ad tres lineas in datis angulis rectæ lineæ ducantur, & data sit proportio rectanguli contenti duabus ductis ad quadratum reliquæ: punctum contingit positione datum solidum locum, hoc est unam ex tribus conicis sectionibus. Et si ad quatuor rectas lineas positione datas in datis angulis lineæ ducantur; & rectanguli duabus ductis contenti ad contentum duabus reliquis proportio data sit: similiter punctum datum coni sectionem positione continget. Si quidem igitur ad duas tantum locus planus ostensus est. Quod si ad plures quam quatuor, punctum continget locos non adhuc cognitos, sed lineas tantum dictas; quales autem sint, vel quam habeant proprietatem, non constat: earum unam, neque primam, & quæ manifestissima videtur, composuerunt ostendentes utilem esse. propositiones autem ipsarum hæ sunt.

Si ab aliquo puncto ad positione datas rectas lineas quinque ducantur rectæ lineæ in datis angulis, & data sit proportio solidi parallelepipedi rectanguli, quod tribus ductis lineis continetur ad solidum parallelepipedum rectangulum, quod continetur reliquis duabus, & data quapiam linea, punctum positione datam lineam continget. Si autem ad sex, & data sit proportio solidi tribus lineis contenti ad solidum, quod tribus reliquis continetur; rursus punctum continget positione datam lineam. Quod si ad plures quam sex, non adhuc habent dicere, an data sit proportio cuiuspiã contenti quatuor lineis ad id quod reliquis continetur, quoniam non est aliquid contentum pluribus quam tribus dimensionibus.

Ou ie vous prie de remarquer en passant, que le scrupule, que faisoient les anciens d'vser des termes de l'Arithmetique en la Geometrie, qui ne pouuoit proceder,

O q   que

que de ce qu'ils ne voyoient pas aſſés clairement leur
rapport, cauſoit beaucoup d'obſcurité, & d'embaras, en
la façon dont ils s'expliquoient. car Pappus pourſuit en
cete ſorte.

*Acquieſcunt autem his, qui paulo ante talia interpretati
ſunt. neque unum aliquo paƈto comprehenſibile ſignificantes
quod his continetur.Licebit autē per coniunƈtas proportiones
hæc, & dicere, & demonſtrare univerſe in diƈtis proportionibus, atque his in hunc modum. Si ab aliquo punƈto ad poſitione datas reƈtas lineas ducantur reƈtæ lineæ in datis angulis, & data ſit proportio coniunƈta ex ea, quam habet una du-
ƈtarum ad unam, & altera ad alteram, & alia ad aliam, & reliqua ad datam lineam, ſi ſint ſeptem; ſi vero oƈto, & reliqua
ad reliquam: punƈtum continget poſitione datās lineas. Et
ſimiliter quotcumque ſint impares vel pares multitudine;
cum hæc, ut dixi, loco ad quatuor lineas reſpondeant, nullum
igitur poſuerunt ita ut linea nota ſit, &c.*

La queſtion donc qui auoit eſté commencée a reſoudre par Euclide, & pourſuiuie par Apollonius, ſans auoir
eſté acheuée par perſonne, eſtoit telle. Ayant trois on
quatre ou plus grand nombre de lignes droites données
par poſition; premierement on demande vn point, duquel on puiſſe tirer autant d'autres lignes droites, vne ſur
chaſcune des données, qui façent auec elles des angles
donnés, & que le reƈtangle contenu en deux de celles,
qui ſeront ainſi tirées d'vn meſme point, ait la proportion donnée auec le quarré de la troiſieſme, s'il n'y en a
que trois; oubien auec le reƈtangle des deux autres, s'il y
en a quatre; oubien, s'il y en a cinq, que le parallelepipede
compoſé de trois ait la proportion donnée auec le parallelepipede

20

quoniam non est aliquid contentum pluribus quam tribus dimensionibus." [33]

Here I beg you to observe in passing that the considerations that forced ancient writers to use arithmetical terms in geometry, thus making it impossible for them to proceed beyond a point where they could see clearly the relation between the two subjects, caused much obscurity and embarrassment, in their attempts at explanation.

Pappus proceeds as follows:

"Acquiescunt autem his, qui paulo ante talia interpretati sunt; neque unum aliquo pacto comprehensibile significantes quod his continetur. Licebit autem per conjunctas proportiones hæc, & dicere & demonstrare universe in dictis proportionibus, atque his in hunc modum. Si ab aliquo puncto ad positione datas rectas lineas ducantur rectæ lineæ in datis angulis, & data sit proportio conjuncta ex ea, quam habet una ductarum ad unam, & altera ad alteram, & alia ad aliam, & reliqua ad datam lineam, si sint septem; si vero octo, & reliqua ad reliquam: punctum continget positione datas lineas. Et similiter quotcumque sint

[33] This may be somewhat freely translated as follows: "The problem of the locus related to three or four lines, about which he (Apollonius) boasts so proudly, giving no credit to the writer who has preceded him, is of this nature: If three straight lines are given in position, and if straight lines be drawn from one and the same point, making given angles with the three given lines; and if there be given the ratio of the rectangle contained by two of the lines so drawn to the square of the other, the point lies on a solid locus given in position, namely, one of the three conic sections.

"Again, if lines be drawn making given angles with four straight lines given in position, and if the rectangle of two of the lines so drawn bears a given ratio to the rectangle of the other two; then, in like manner, the point lies on a conic section given in position. It has been shown that to only two lines there corresponds a plane locus. But if there be given more than four lines, the point generates loci not known up to the present time (that is, impossible to determine by common methods), but merely called 'lines'. It is not clear what they are, or what their properties. One of them, not the first but the most manifest, has been examined, and this has proved to be helpful. (Paul Tannery, in the *Oeuvres de Descartes*, differs with Descartes in his translation of Pappus. He translates as follows: Et on n'a fait la synthèse d' aucune de ces lignes, ni montré qu'elle servit pour ces lieux, pas même pour celle qui semblerait la première et la plus indiquée.) These, however, are the propositions concerning them.

"If from any point straight lines be drawn making given angles with five straight lines given in position, and if the solid rectangular parallelepiped contained by three of the lines so drawn bears a given ratio to the solid rectangular parallelepiped contained by the other two and any given line whatever, the point lies on a 'line' given in position. Again, if there be six lines, and if the solid contained by three of the lines bears a given ratio to the solid contained by the other three lines, the point also lies on a 'line' given in position. But if there be more than six lines, we cannot say whether a ratio of something contained by four lines is given to that which is contained by the rest, since there is no figure of more than three dimensions."

21

impares vel pares multitudine, cum hæc, ut dixi, loco ad quatuor lineas respondeant, nullum igitur posuerunt ita ut linea nota sit, &c.[34]

The question, then, the solution of which was begun by Euclid and carried farther by Apollonius, but was completed by no one, is this:

Having three, four or more lines given in position, it is first required to find a point from which as many other lines may be drawn, each making a given angle with one of the given lines, so that the rectangle of two of the lines so drawn shall bear a given ratio to the square of the third (if there be only three); or to the rectangle of the other two (if there be four), or again, that the parallelepiped[35] constructed upon three shall bear a given ratio to that upon the other two and any given line (if there be five), or to the parallelepiped upon the other three (if there be six); or (if there be seven) that the product obtained by multiplying four of them together shall bear a given ratio to the product of the other three, or (if there be eight) that the product of four of them shall bear a given ratio to the product of the other four. Thus the question admits of extension to any number of lines.

Then, since there is always an infinite number of different points satisfying these requirements, it is also required to discover and trace the curve containing all such points.[36] Pappus says that when there are only three or four lines given, this line is one of the three conic sections, but he does not undertake to determine, describe, or explain the nature of the line required[37] when the question involves a greater number of lines. He only adds that the ancients recognized one of them which they had shown to be useful, and which seemed the sim-

[34] This rather obscure passage may be translated as follows: "For in this are agreed those who formerly interpreted these things (that the dimensions of a figure cannot exceed three) in that they maintain that a figure that is contained by these lines is not comprehensible in any way. This is permissible, however, both to say and to demonstrate generally by this kind of proportion, and in this manner: If from any point straight lines be drawn making given angles with straight lines given in position; and if there be given a ratio compounded of them, that is the ratio that one of the lines drawn has to one, the second has to a second, the third to a third, and so on to the given line if there be seven lines, or, if there be eight lines, of the last to a last, the point lies on the lines that are given in position. And similarly, whatever may be the odd or even number, since these, as I have said, correspond in position to the four lines; therefore they have not set forth any method so that a line may be known." The meaning of the passage appears from that which follows in the text.

[35] That is, continued product.

[36] It is here that the essential feature of the work of Descartes may be said to begin.

[37] See line 19 on the opposite page.

lelepipede compoſé des deux qui reſtent, & d'vne autre
ligne donnée. Ou s'il y en a fix, que le parallelepipede
côpoſé de trois ait la proportion donnée auec le parallelepipede des trois autres. Ou s'il y en a ſept, que ce qui ſe
produiſt lorſqu'on en multiplie quatre l'vne par l'autre,
ait la raiſon donnée auec ce qui ſe produiſt par la multiplication des trois autres, & encore d'vne autre ligne
donnée; Ou s'il y en a huit, que le produit de la multiplication de quatre ait la proportion donnée auec le produit des quatre autres. Et ainſi cete queſtion ſe peut
eſtendre a tout autre nombre de lignes. Puis a cauſe qu'il
y a touſiours vne infinité de diuers poins qui peuuent ſa-
tisfaire a ce qui eſt icy demandé, il eſt auſſy requis de
connoiſtre, & de tracer la ligne, dans laquelle ils doiuent
tous ſe trouuer. & Pappus dit que lorſqu'il n'y a que
trois ou quatre lignes droites données, c'eſt en vne des
trois ſections coniques. mais il n'entreprend point de la
determiner, ny de la deſcrire. non plus que d'expliquer celles ou tous ces poins ſe doiuent trouuer, lorſque
la queſtion eſt propoſée en vn plus grand nombre de lignes. Seulement il aiouſte que les anciens en auoient
imaginé vne qu'ils monſtroient y eſtre vtile , mais qui
ſembloit la plus manifeſte, & qui n'eſtoit pas toutefois la
premiere. Ce qui m'a donné occaſion d'eſſayer ſi par la
methode dont ie me ſers on peut aller auſſy loin qu'ils
ont eſté.

Et premierement i'ay connu que cete queſtion n'eſtant
propoſée qu'en trois, ou quatre, ou cinq lignes , on peut
touſiours trouuer les poins cherchés par la Geometrie
ſimple; c'eſt a dire en ne ſe ſeruant que de la reigle & du

Reſponſe
à la que-
ſtion de
Pappus.

Qq 2   compas,

compas, ny ne faisant autre chose, que ce qui a desia esté
dit; excepté seulement lorsqu'il y a cinq lignes données,
si elles sont toutes paralleles. Auquel cas, comme aussy
lorsque la question est proposée en six, ou 7, ou 8, ou 9
lignes, on peut tousiours trouuer les poins cherchés par
la Geometrie des solides; c'est a dire en y employant
quelqu'vne des trois sections coniques. Excepté seulement lorsqu'il y a neuf lignes données, si elles sont toutes
paralleles. Auquel cas derechef, & encore en 10, 11, 12,
ou 13 lignes on peut trouuer les poins cherchés par le
moyen d'vne ligne courbe qui soit d'vn degré plus composée que les sections coniques. Excepté en treize si elles sont toutes paralleles, auquel cas, & en quatorze, 15,
16, & 17 il y faudra employer vne ligne courbe encore
d'vn degré plus composée que la precedente & ainsi
a l'infini.

Puis iay trouué aussy, que lorsqu'il n'y a que trois ou
quatre lignes données, les poins cherchés se rencontrent
tous , non seulement en l'vne des trois sections coniques, mais quelquefois aussy en la circonference d'vn
cercle, ou en vne ligne droite. Et que lorsqu'il y en a
cinq, ou six, ou sept, ou huit, tous ces poins se rencontrent en quelque vne des lignes, qui sont d'vn degré plus
composées que les sections coniques , & il est impossible
d'en imaginer aucune qui ne soit vtile a cete question;
mais ils peuuent aussy derechef se rencontrer en vne section conique, ou en vn cercle, ou en vne ligne droite.
Et s'il y en a neuf, ou 10, ou 11, ou 12, ces poins se rencontrent en vne ligne, qui ne peut estre que d'vn degré
plus composée que les precedentes; mais toutes celles
qui

plest, and yet was not the most important.[38] This led me to try to find out whether, by my own method, I could go as far as they had gone.[39]

First, I discovered that if the question be proposed for only three, four, or five lines, the required points can be found by elementary geometry, that is, by the use of the ruler and compasses only, and the application of those principles that I have already explained, except in the case of five parallel lines. In this case, and in the cases where there are six, seven, eight, or nine given lines, the required points can always be found by means of the geometry of solid loci,[40] that is, by using some one of the three conic sections. Here, again, there is an exception in the case of nine parallel lines. For this and the cases of ten, eleven, twelve, or thirteen given lines, the required points may be found by means of a curve of degree next higher than that of the conic sections. Again, the case of thirteen parallel lines must be excluded, for which, as well as for the cases of fourteen, fifteen, sixteen, and seventeen lines, a curve of degree next higher than the preceding must be used; and so on indefinitely.

Next, I have found that when only three or four lines are given, the required points lie not only all on one of the conic sections but sometimes on the circumference of a circle or even on a straight line.[41]

When there are five, six, seven, or eight lines, the required points lie on a curve of degree next higher than the conic sections, and it is impossible to imagine such a curve that may not satisfy the conditions of the problem; but the required points may possibly lie on a conic section, a circle, or a straight line. If there are nine, ten, eleven, or twelve lines, the required curve is only one degree higher than the preceding, but any such curve may meet the requirements, and so on to infinity.

[38] See lines 5-10 from the foot of page 23.

.[39] Descartes gives here a brief summary of his solution, which he amplifies later.

[40] This term was commonly applied by mathematicians of the seventeenth century to the three conic sections, while the straight line and circle were called plane loci, and other curves linear loci. See Fermat, *Isagoge ad Locos Planos et Solidos,* Toulouse, 1679.

[41] Degenerate or limiting forms of the conic sections.

25

Finally, the first and simplest curve after the conic sections is the one generated by the intersection of a parabola with a straight line in a way to be described presently.

I believe that I have in this way completely accomplished what Pappus tells us the ancients sought to do, and I will try to give the demonstration in a few words, for I am already wearied by so much writing.

Let AB, AD, EF, GH, ... be any number of straight lines given in position,[42] and let it be required to find a point C, from which straight lines CB, CD, CF, CH, ... can be drawn, making given angles CBA, CDA, CFE, CHG, ... respectively, with the given lines, and

---

[42] It should be noted that these lines are given in position but not in length. They thus become lines of reference or coördinate axes, and accordingly they play a very important part in the development of analytic geometry. In this connection we may quote as follows: "Among the predecessors of Descartes we reckon, besides Apollonius, especially Vieta, Oresme, Cavalieri, Roberval, and Fermat, the last the most distinguished in this field; but nowhere, even by Fermat, had any attempt been made to refer several curves of different orders simultaneously to one system of coördinates, which at most possessed special significance for one of the curves. It is exactly this thing which Descartes systematically accomplished." Karl Fink, *A Brief History of Mathematics,* trans. by Beman and Smith, Chicago, 1903, p. 229.

Heath calls attention to the fact that "the essential difference between the Greek and the modern method is that the Greeks did not direct their efforts to making the fixed lines of a figure as few as possible, but rather to expressing their equations between areas in as short and simple a form as possible." For further discussion see D. E. Smith, *History of Mathematics,* Boston, 1923-25, Vol. II, pp. 316-331 (hereafter referred to as Smith).

qui font d'vn degré plus compofées y peuuent feruir, &
ainfi a l'infini.

Au refte la premiere, & la plus fimple de toutes aprés
les fections coniques, eft celle qu'on peut defcrire par
l'interfection d'vne Parabole, & d'vne ligne droite, en la
façon qui fera tantoft expliquée. En forte que ie penfe
auoir entierement fatisfait a ceque Pappus nous dit auoir
efté chetché en cecy par les anciens. & ie tafcheray d'en
mettre la demonftration en peu de mots. car il m'ennuie
defia d'en tant efcrire.

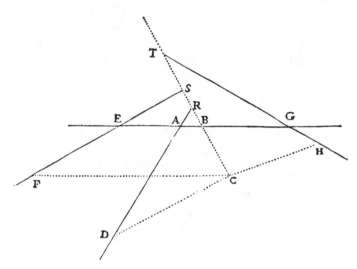

Soient A B, A D, E F, G H, &c. plufieurs lignes donnees par pofition, & qu'il faille trouuer vn point, comme
C, duquel ayant tiré d'autres lignes droites fur les données, comme C B, C D, C F, & C H, en forte que les
angles C B A, C D A, C F E, C H G, &c. foient donnés,

Q q 3   &

& que ce qui eſt produit par la multiplication d'vne partie de ces lignes, ſoit eſgal a ce qui eſt produit par la multiplication des autres, oubien qu'ils ayent quelque autre
proportion donnée, car cela ne rend point la queſtion
plus difficile.

Commĕt
on doit
poſer les
termes
pour ve-
nir à l'E-
quation
en cet
exemple.

Premierement ie ſuppoſe la choſe comme deſia faite,
& pour me demeſler de la côfuſion de toutes ces lignes,
ie conſidere l'vne des données, & l'vne de celles qu'il
faut trouuer, par exemple A B, & C B, comme les prin-
cipales, & auſquelles ie taſche de rapporter ainſi toutes
les autres. Que le ſegment de la ligne A B, qui eſt entre
les poins A & B, ſoit nommé x. & que B C ſoit nommé
y. & que toutes les autres lignes données ſoient prolon-
gées, iuſques a ce qu'elles couppent ces deux, auſſy pro-
longées s'il eſt beſoin, & ſi elles ne leur ſont point paralleles. comme vous voyes icy qu'elles couppent la ligne
A B aux poins A, E, G, & B C aux poins R, S, T. Puis a
cauſe que tous les angles du triangle A R B ſont donnés,
la proportion, qui eſt entre les coſtés A B, & B R, eſt auſſy donnée, & ie la poſe comme de z à b, de façon qu' A B
eſtant x, R B ſera $\frac{bx}{z}$, & la toute C R ſera $y + \frac{bx}{z}$, à cauſe
que le point B tombe entre C & R; car ſi R tomboit en-
tre C & B, C R ſeroit $y - \frac{bx}{z}$; & ſi C tomboit entre B & R,
C R ſeroit $- y + \frac{bx}{z}$. Tout de meſme les trois angles
du triangle D R C ſont donnés, & par conſequent auſſy
la proportion qui eſt entre les coſtés C R, & C D, que ie
poſe comme de z à c: de façon que C R eſtant $y + \frac{bx}{z}$,

CD

such that the product of certain of them is equal to the product of the rest, or at least such that these two products shall have a given ratio, for this condition does not make the problem any more difficult.

First, I suppose the thing done, and since so many lines are confusing, I may simplify matters by considering one of the given lines and one of those to be drawn (as, for example, AB and BC) as the principal lines, to which I shall try to refer all the others. Call the segment of the line AB between A and B, *x,* and call BC, *y.* Produce all the other given lines to meet these two (also produced if necessary) provided none is parallel to either of the principal lines. Thus, in the figure, the given lines cut AB in the points A, E, G, and cut BC in the points R, S, T.

Now, since all the angles of the triangle ARB are known,[43] the ratio between the sides AB and BR is known.[44] If we let $AB : BR = z : b$, since $AB = x$, we have $RB = \dfrac{bx}{z}$; and since B lies between C and R [45], we have $CR = y + \dfrac{bx}{z}$. (When R lies between C and B, CR is equal to $y - \dfrac{bx}{z}$, and when C lies between B and R, CR is equal to $-y + \dfrac{bx}{z}$)

Again, the three angles of the triangle DRC are known,[46] and therefore the ratio between the sides CR and CD is determined. Calling this ratio $z : c$, since $CR = y + \dfrac{bx}{z}$, we have $CD = \dfrac{cy}{z} + \dfrac{b.x}{z^2}$. Then, since

[43] Since BC cuts AB and AD under given angles.
[44] Since the ratio of the sines of the opposite angles is known.
[45] In this particular figure, of course.
[46] Since CB and CD cut AD under given angles.

the lines AB, AD, and EF are given in position, the distance from A to E is known. If we call this distance k, then $EB = k + x$; although $EB = k - x$ when B lies between E and A, and $E = -k + x$ when E lies between A and B. Now the angles of the triangle ESB being given, the ratio of BE to BS is known. We may call this ratio $z : d$.

Then $BS = \dfrac{dk + dx}{z}$ and $CS = \dfrac{zy + dk + dx}{z}$.[47] When S lies between B and C we have $CS = \dfrac{zy - dk - dx}{z}$, and when C lies between B and S we have $CS = \dfrac{-zy + dk + dx}{z}$. The angles of the triangle FSC are known, and hence, also the ratio of CS to CF, or $z : e$. Therefore, $CF = \dfrac{ezy + dek + dex}{z^2}$. Likewise, AG or l is given, and $BG = l - x$. Also, in triangle BGT, the ratio of BG to BT, or $z : f$, is known. Therefore, $BT = \dfrac{fl - fx}{z}$ and $CT = \dfrac{zy + fl - fx}{z}$. In triangle TCH, the ratio of TC to CH, or $z : g$, is known,[48] whence $CH = \dfrac{gzy + fgl - fgx}{z^2}$.

[47] We have
$$CS = y + BS$$
$$= y + \frac{dk + dx}{z}$$
$$= \frac{zy + dk + dx}{z},$$

and similarly for the other cases considered below.

The translation covers the first eight lines on the original page 312 (page 32 of this edition.

[48] It should be noted that each ratio assumed has z as antecedent.

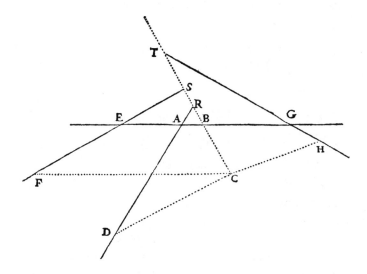

CD fera $\frac{cy}{z} + \frac{bcx}{zz}$. Aprés cela pourceque les lignes AB,
AD, & EF font données par pofition, la diftance qui eft
entre les poins A & E eft auffy donnée, & fi on la nomme K, on aura EB efgal a $k + x$; mais ce feroit $k - x$, fi
le point B tomboit entre E & A; & $- k + x$, fi E tomboit
entre A & B. Et pourceque les angles du triangle ESB
font tous donnés, la proportion de BE a BS eft auffy
donnée, & ie la pofe comme z à d, fibienque BS eft
$\frac{dk + dx}{z}$, & la toute CS eft $\frac{zy + dk + dx}{z}$; mais ce feroit
$\frac{zy - dk - dx}{z}$, fi le point S tomboit entre B & C; & ce feroit
$\frac{- zy + dk + dx}{z}$, fi C tomboit entre B & S. De plus les
trois angles du triangle FSC font donnés, & en fuite la
pro-

proportion de C S à C F, qui ſoit comme de χ à e, & la
toute C F ſera $\frac{ezy \pm dek \pm dex}{zz}$. En meſme façon A G
que ie nomme l eſt donnée, & B G eſt $l -- x$, & a cauſe
du triangle B G T la proportion de B G à B T eſt auſſy
donnée, qui ſoit comme de χ à f. & B T ſera $\frac{fl--fx}{\chi}$, &
C T ∞ $\frac{\chi y \pm fl--fx}{z}$. Puis derechef la proportion de T C à
C H eſt donnée, a cauſe du triangle T C H, & la poſant
comme de χ à g, on aura C H ∞ $\frac{\pm gzy \pm fgl--fgx}{zz}$.

   Et ainſi vous voyés, qu'en tel nombre de lignes données par poſition qu'on puiſſe auoir, toutes les lignes tirées deſſus du point C a angles donnés ſuiuant la teneur
de la queſtion, ſe peuuent touſiours exprimer chaſcune
par trois termes; dont l'vn eſt compoſé de la quantité inconnue y, multipliée, ou diuiſee par quelque autre
connue; & l'autre de la quantité inconnue x, auſſy multipliée ou diuiſée par quelque autre connuë, & le troſieſ-
me d'vne quantité toute connuë. Excepté ſeulement ſi
elles ſont paralleles; oubien a la ligne A B, auquel cas le
terme compoſé de la quantité x ſera nul ; oubien a la ligne C B, auquel cas celuy qui eſt compoſé de la quantité
y ſera nul; ainſi qu'il eſt trop manifeſte pour que ie m'areſte a l'expliquer. Et pour les ſignes $\pm$, & $--$, qui ſe ioi-
gnent à ces termes, ils peuuent eſtre changés en toutes
les façons imaginables.

   Puis vous voyés auſſy, que multipliant pluſieurs de
ces lignes l'vne par l'autre, les quantités x & y, qui ſe
trouuent dans le produit, n'y peuuent auoir que chaſcune autant de dimenſions, qu'il y a eu de lignes, a l'explication

And thus you see that, no matter how many lines are given in position, the length of any such line through C making given angles with these lines can always be expressed by three terms, one of which consists of the unknown quantity y multiplied or divided by some known quantity; another consisting of the unknown quantity x multiplied or divided by some other known quantity; and the third consisting of a known quantity.[49] An exception must be made in the case where the given lines are parallel either to AB (when the term containing x vanishes), or to CB (when the term containing y vanishes). This case is too simple to require further explanation.[50] The signs of the terms may be either + or — in every conceivable combination.[51]

You also see that in the product of any number of these lines the degree of any term containing x or y will not be greater than the number of lines (expressed by means of x and y) whose product is found. Thus, no term will be of degree higher than the second if two lines be multiplied together, nor of degree higher than the third, if there be three lines, and so on to infinity.

[49] That is, an expression of the form $ax + by + c$, where a, b, c, are any real positive or negative quantities, integral or fractional (not zero, since this exception is considered later).

[50] The following problem will serve as a very simple illustration: Given three parallel lines AB, CD, EF, so placed that AB is distant 4 units from CD, and CD is distant 3 units from EF; required to find a point P such that if PL, PM, PN

be drawn through P, making angles of 90°, 45°, 30°, respectively, with the parallels. Then $\overline{PM}^2 = PL \cdot PN$.

Let $PR = y$, then $PN = 2y$, $PM = \sqrt{2}(y+3)$, $PL = y+7$. If $\overline{PM}^2 = PN \cdot PL$, we have $\left[\sqrt{2}(y+3) \right]^2 = 2y(y+7)$, whence $y = 9$. Therefore, the point P lies on the line XY parallel to EF and at a distance of 9 units from it. Cf. Rabuel, p. 79.

[51] Depending, of course, upon the relative positions of the given lines.

Furthermore, to determine the point C, but one condition is needed, namely, that the product of a certain number of lines shall be equal to, or (what is quite as simple), shall bear a given ratio to the product of certain other lines. Since this condition can be expressed by a single equation in two unknown quantities,[52] we may give any value we please to either x or y and find the value of the other from this equation. It is obvious that when not more than five lines are given, the quantity x, which is not used to express the first of the lines can never be of degree higher than the second.[53]

Assigning a value to y, we have $x^2 = \pm ax \pm b^2$, and therefore x can be found with ruler and compasses, by a method already explained.[54] If then we should take successively an infinite number of different values for the line y, we should obtain an infinite number of values for the line x, and therefore an infinity of different points, such as C, by means of which the required curve could be drawn.

This method can be used when the problem concerns six or more lines, if some of them are parallel to either AB or BC, in which case

[52] That is, an indeterminate equation. "De plus, à cause que pour determiner le point C, il n'y a qu'une seule condition qui soit requise, à sçavoir que ce qui est produit par la multiplication d'un certain nombre de ces lignes soit égal, ou (ce qui n'est de rien plus mal-aisé) ait la proportion donnee, à ce qui est produit par la multiplication des autres; on peut prendre à discretion l'une des deux quantitez inconnuës x ou y, & chercher l'autre par cette Equation." Such variations in the texts of different editions are of no moment, but are occasionally introduced as matters of interest.

[53] Since the product of three lines bears a given ratio to the product of two others and a given line, no term can be of higher degree than the third, and therefore, than the second in x.

[54] See pages 13, et seq.

cation defquelles elles feruent, qui ont efté ainfi multipliées: enforte qu'elles n'auront iamais plus de deux dimenfions, en ce qui ne fera produit que par la multiplication de deux lignes; ny plus de trois, en ce qui ne fera
produit que par la multiplication de trois, & ainfi a l'infini.

De plus, a caufe que pour determiner le point C, il Commēt on trouue
n'y a qu'vne feule condition qui foit requife, à fçauoir que ce probleſ-
que ce qui eft produit par la multiplication d'vn certain me eſt
nombre de ces lignes foit efgal, ou (cequi n'eft de rien plan, lorſ-
plus malayfé) ait la proportion donnée, à ce qui eft pro- qu'il n'eſt point
duit par la multiplication des autres; on peut prendre a propofé en plus de
difcretion l'vne des deux quantités inconnues x ou y, & 5 lignes.
chercher l'autre par cete Equation. en laquelle il eft euident que lorfque la queftion n'eft point propofée en plus
de cinq lignes, la quantité x qui ne fert point a l'expref-
fion de la premiere peut toufiours n'y auoir que deux dimenfions. de façon que prenant vne quantité connuë
pour y, il ne reftera que $xx \infty + $ ou $-- ax +$ ou $-- bb$. &
ainfi on pourra trouuer la quantité x auec la reigle & le
compas, en la façon tantoft expliquée. Mefme prenant
fucceffiuement infinies diuerfes grandeurs pour la ligne
y, on en trouuera auffy infinies pour la ligne x, & ainfi on
aura vne infinité de diuers poins, tels que celuy qui eft
marqué C, par le moyen defquels on defcrira la ligne
courbe demandée.

Il fe peut faire auffy, la queftion eftant propofée en fix,
ou plus grand nombre de lignes; s'il y en a entre les données, qui foient paralleles a B A, ou B C, que l'vne des
deux quantités x ou y n'ait que deux dimenfions en

R r l'Equal'Equation, & ainſi qu'on puiſſe trouuuer le point C auec la reigle & le compas. Mais au contraire ſi elles ſout toutes paralleles , encore que la queſtion ne ſoit propoſée qu'en cinq lignes, ce point C ne pourra ainſi eſtre trouué, a cauſe que la quantité *x* ne ſe trouuant point en toute l'Equation, il ne ſera plus permis de prendre vne quantité connuë pour celle qui eſt nommée *y* , mais ce ſera elle qu'il faudra chercher. Et pource quelle aura trois dimenſions, on ne la pourra trouuer qu'en tirant la racine d'vne Equation cubique. cequi ne ſe peut generalement faire ſans qu'on y employe pour le moins vne ſection conique. Et encore qu'il y ait iuſques a neuf lignes données, pourvûqu'elles ne ſoient point toutes paralleles, on peut touſiours faire que l'Equation ne monte que iuſques au quarré de quarré. au moyen dequoy on la peut auſſy touſiours reſoudre par les ſections coniques, en la façon que i'expliqueray cy aprés. Et encore qu'il y en ait iuſques a treize , on peut touſiours faire qu'elle ne monte que iuſques au quarré de cube. en ſuite de quoy on la peut reſoudre par le moyen d'vne ligne , qui n'eſt que d'vn degré plus compoſée que les ſections coniques, en la façon que i'expliqueray auſſy cy aprés. Et cecy eſt la premiere partie de ceque i'auois icy a demonſtrer ; mais auant que ie paſſe a la ſeconde il eſt beſoin que ie die quelque choſe en general de la nature des lignes courbes.

LA

either x or y will be of only the second degree in the equation, so that the point C can be found with ruler and compasses.

On the other hand, if the given lines are all parallel even though a question should be proposed involving only five lines, the point C cannot be found in this way. For, since the quantity x does not occur at all in the equation, it is no longer allowable to give a known value to y. It is then necessary to find the value of y.[55] And since the term in y will now be of the third degree, its value can be found only by finding the root of a cubic equation, which cannot in general be done without the use of one of the conic sections.[56]

And furthermore, if not more than nine lines are given, not all of them being parallel, the equation can always be so expressed as to be of degree not higher than the fourth. Such equations can always be solved by means of the conic sections in a way that I shall presently explain.[57]

Again, if there are not more than thirteen lines, an equation of degree not higher than the sixth can be employed, which admits of solution by means of a curve just one degree higher than the conic sections by a method to be explained presently.[58]

This completes the first part of what I have to demonstrate here, but it is necessary, before passing to the second part, to make some general statements concerning the nature of curved lines.

[55] That is, to solve the equation for y.
[56] See page 84.
[57] See page 107.
[58] This line of reasoning may be extended indefinitely. Briefly, it means that for every two lines introduced the equation becomes one degree higher and the curve becomes correspondingly more complex.

BOOK SECOND

Geometry

BOOK II

On the Nature of Curved Lines

THE ancients were familiar with the fact that the problems of geometry may be divided into three classes, namely, plane, solid, and linear problems.[59] This is equivalent to saying that some problems require only circles and straight lines for their construction, while others require a conic section and still others require more complex curves.[60] I am surprised, however, that they did not go further, and distinguish between different degrees of these more complex curves, nor do I see why they called the latter mechanical, rather than geometrical.[61] If we say that they are called mechanical because some sort of instrument[62] has to be used to describe them, then we must, to be consistent,

---

[50] Cf. Pappus, Vol. I, p. 55, Proposition 5, Book III: "The ancients considered three classes of geometric problems, which they called plane, solid, and linear. Those which can be solved by means of straight lines and circumferences of circles are called plane problems, since the lines or curves by which they are solved have their origin in a plane. But problems whose solutions are obtained by the use of one or more of the conic sections are called solid problems, for the surfaces of solid figures (conical surfaces) have to be used. There remains a third class which is called linear because other 'lines' than those I have just described, having diverse and more involved origins, are required for their construction. Such lines are the spirals, the quadratrix, the conchoid, and the cissoid, all of which have many important properties." See also Pappus, Vol. I, p. 271.

[60] Rabuel (p. 92) suggests dividing problems into classes, the first class to include all problems that can be constructed by means of straight lines, that is, curves whose equations are of the first degree; the second, those that require curves whose equations are of the second degree, namely, the circle and the conic sections, and so on.

[61] Cf. *Encyclopédie ou Dictionnaire Raisonné des Sciences, des Arts et des Metiers, par une Société de gens de lettres, mis en ordre et publiées par M. Diderot, et quant à la Partie Mathematique par M. d'Alembert*, Lausanne and Berne, 1780. In substance as follows: "*Mechanical* is a mathematical term designating a construction not geometric, that is, that cannot be accomplished by geometric curves. Such are constructions depending upon the quadrature of the circle.

The term, mechanical curve, was used by Descartes to designate a curve that cannot be expressed by an algebraic equation." Leibniz and others call them transcendental.

[62] "Machine."

40

LA
GEOMETRIE.
LIVRE SECOND.

De la nature des lignes courbes.

L Es anciens ont fort bien remarqué, qu'entre les Problefmes de Geometrie, les vns font plans, les autres folides, & les autres lineaires, c'eft a dire, que les vns peuuent eftre conftruits, en ne traçant que des lignes droites, & des cercles; au lieu que les autres ne le peuuent eftre, qu'on n'y employe pour le moins quelque fection conique; ni enfin les autres, qu'on n'y employe quelque autre ligne plus compofée. Mais ie m'eftonne de ce qu'ils n'ont point outre cela diftingué diuers degrés entre ces lignes plus compofées, & ie ne fçaurois comprendre pourquoy ils les ont nommées mechaniques, plutoft que Geometriques. Car de dire que ç'ait efté, a caufe qu'il eft befoin de fe feruir de quelque machine pour les defcrire, il faudroit reietter par mefme raifon les cercles & les lignes droites; vû qu'on ne les defcrit fur le papier qu'auec vn compas, & vne reigle, qu'on peut auffy nommer des machines. Ce n'eft pas non plus, a caufe que les inftrumens, qui feruent a les tracer, eftant plus compofés que la reigle & le compas, ne peuuent eftre fi iuftes; car il faudroit pour cete raifon les reietter des Mechaniques, où la iufteffe des ouurages qui fortent de la main eft defirée; plutoft que de la Geometrie, ou c'eft feulement la iufteffe du raifonnemēt qu'on recher-

(marginal note:) Quelles font les lignes courbes qu'on peut receuoir en Geometrie.

R r 2   che,

che, & qui peut fans doute eftre auffy parfaite touchant
ces lignes, que touchant les autres. Ie ne diray pas auffy,
que ce foit a caufe qu'ils n'ont pas voulu augmenter le
nombre de leurs demandes., & qu'ils fe font contentés
qu'on leur accordaft, qu'ils puffent ioindre deux poins
donnés par vne ligne droite , & defcrire vn cercle d'vn
centre donné, qui paffaft par vn point donné. car ils n'ont
point fait de fcrupule de fuppofer outre cela, pour traiter
des fections coniques , qu'on puft coupper tout cóne
donné par vn plan donné. & il n'eft befoin de rien fuppofer pour traçer toutes les lignes courbes, que ie pre-
tens icy d'introduire; finon que deux ou plufieurs lignes
puiffent eftre meuës l'vne par l'autre, & que leurs inter-
fections en marquent d'autres ; ce qui ne me paroift en
rien plus difficile. Il eft vray qu'ils n'ont pas auffy entierement receu les fections coniques en leur Geometrie,
& ie ne veux pas entreprendre de changer les noms qui
ont efté approuués par l'vfage; mais il eft, ce me femble,
tres clair, que prenant comme on fait pour Geometrique ce qui eft precis & exact , & pour Mechanique
ce qui ne l'eft pas ; & confiderant la Geometrie comme
vne fcience, qui enfeigne generalement a connoiftre les
mefures de tous les cors, on n'en doit pas plutoft exclure
les lignes les plus compofées que les plus fimples, pourvû
qu'on les puiffe imaginer eftre defcrites par vn mouuement continu, ou par plufieurs qui s'entrefuiuent & dont
les derniers foient entierement reglés par ceux qui les
precedent. car par ce moyen on peut toufiours auoir
vne connoiffance exacte de leur mefure. Mais peuteftre
que ce qui a empefché les anciens Geometres de reçeuoir

reject circles and straight lines, since these cannot be described on paper without the use of compasses and a ruler, which may also be termed instruments. It is not because the other instruments, being more complicated than the ruler and compasses, are therefore less accurate, for if this were so they would have to be excluded from mechanics, in which accuracy of construction is even more important than in geometry. In the latter, exactness of reasoning alone[63] is sought, and this can surely be as thorough with reference to such lines as to simpler ones.[64] I cannot believe, either, that it was because they did not wish to make more than two postulates, namely, (1) a straight line can be drawn between any two points, and (2) about a given center a circle can be described passing through a given point. In their treatment of the conic sections they did not hesitate to introduce the assumption that any given cone can be cut by a given plane. Now to treat all the curves which I mean to introduce here, only one additional assumption is necessary, namely, two or more lines can be moved, one upon the other, determining by their intersection other curves. This seems to me in no way more difficult.[65]

It is true that the conic sections were never freely received into ancient geometry,[66] and I do not care to undertake to change names confirmed by usage; nevertheless, it seems very clear to me that if we make the usual assumption that geometry is precise and exact, while mechanics is not;[67] and if we think of geometry as the science which furnishes a general knowledge of the measurement of all bodies, then we have no more right to exclude the more complex curves than the simpler ones, provided they can be conceived of as described by a continuous motion or by several successive motions, each motion being completely determined by those which precede; for in this way an exact knowledge of the magnitude of each is always obtainable.

[63] An interesting question of modern education is here raised, namely, to what extent we should insist upon accuracy of construction even in elementary geometry.
[64] Not only ancient writers but later ones, up to the time of Descartes, made the same distinction; for example, Vieta. Descartes's view has been universally accepted since his time.
[65] That is, in no way less obvious than the other postulates.
[66] Because the ancients did not believe that the so-called constructions of the conic sections on a plane surface could be exact.
[67] Since it is not possible to construct an ideal line, plane, and so on.

Probably the real explanation of the refusal of ancient geometers to accept curves more complex than the conic sections lies in the fact that the first curves to which their attention was attracted happened to be the spiral,[68] the quadratrix,[69] and similar curves, which really do belong only to mechanics, and are not among those curves that I think should be included here, since they must be conceived of as described by two separate movements whose relation does not admit of exact determination. Yet they afterwards examined the conchoid,[70] the cissoid,[71] and a few others which should be accepted; but not knowing much about their properties they took no more account of these than of the others. Again, it may have been that, knowing as they did only a little about the conic sections,[72] and being still ignorant of many of the possibilities of the ruler and compasses, they dared not yet attack a matter of still greater difficulty. I hope that hereafter those who are clever enough to make use of the geometric methods herein suggested will find no great difficulty in applying them to plane or solid problems. I therefore think it proper to suggest to such a more extended line of investigation which will furnish abundant opportunities for practice.

Consider the lines AB, AD, AF, and so forth (page 46), which we may suppose to be described by means of the instrument YZ. This instrument consists of several rulers hinged together in such a way that YZ being placed along the line AN the angle XYZ can be increased or decreased in size, and when its sides are together the points B, C, D, E, F, G, H, all coincide with A; but as the size of the angle is increased,

[68] See Heath, *History of Greek Mathematics* (hereafter referred to as Heath). Cambridge, 2 vols., 1921. Also Cantor, *Vorlesungen über Geschichte der Mathematik*, Leipzig, Vol. I (2), p. 263, and Vol. II (1), pp. 765 and 781 (hereafter referred to as Cantor).

[69] See Heath, I, 225; Smith, Vol. II, pp. 300, 305.

[70] See Heath, I, 235, 238; Smith, Vol. II, p. 298.

[71] See Heath, I, 264; Smith, Vol. II, p. 314.

[72] They really knew much more than would be inferred from this statement. In this connection, see Taylor, *Ancient and Modern Geometry of Conics*, Cambridge, 1881.

uoir celles qui eſtoient plus compoſées que les ſeƈtions
coniques, c'eſt que les premieres qu'ils ont conſiderées,
ayant par haſard eſté la Spirale, la Quadratrice, & ſemblables, qui n'appartienent veritablement qu'aux Mechaniques, & ne ſont point du nombre de celles que ie
penſe deuoir icy eſtre receues, a cauſe qu'on les imagine
deſcrites par deux monuemens ſeparés, & qui n'ont entre eux aucun raport qu'on puiſſe meſurer exaƈtement,
bienqu'ils ayent aprés examiné la Conchoide, la Ciſſoi-
de, & quelque peu d'autres qui en ſont, toutefois a cauſe qu'ils n'ont peuteſtre pas aſſés remarqué leurs proprietés, ils n'en ont pas fait plus d'eſtat que des premieres. Oubien c'eſt que voyant, qu'ils ne connoiſſoient
encore, que peu de choſes touchant les ſeƈtions coniques, & qu'il leur en reſtoit meſme beaucoup, touchant
ce qui ſe peut faire auec la reigle & le compas, qu'ils
ignoroient, ils ont creu ne deuoir point entamer de matiere plus difficile. Mais pourceque i'eſpere que d'orena-
uant ceux qui auront l'adreſſe de ſe ſeruir du calcul Geometrique icy propoſé, ne trouueront pas aſſés dequoy
s'areſter touchant les probleſmes plans, ou ſolides; ie
croy qu'il eſt a propos que ie les inuite a d'autres re-
cherches, où ils ne manqueront iamais d'exercice.

Voyés les lignes A B, A D, A F, & ſemblables que ie
ſuppoſe auoir eſté deſcrites par l'ayde de l'inſtrument
Y Z, qui eſt compoſé de pluſieurs reigles tellement iointes, que celle qui eſt marquée Y Z eſtant areſtée ſur la
ligne A N, on peut ouurir & fermer l'angle X Y Z; & que
lorſqu'il eſt tout fermé, les poins B, C, D, F, G, H ſont
tous aſſemblés au point A ; mais qu'a meſure qu'on

R r 3 l'ouure,

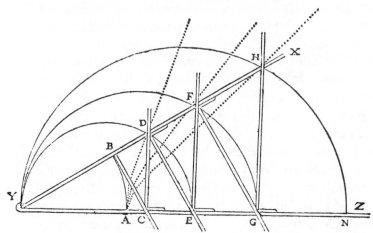

l'ouure, la reigle B C, qui eft iointe a angles droits auec
X Y au point B, pouffe vers Z la reigle C D, qui coule
fur Y Z en faifant toufiours des angles droits auec elle, &
C D pouffe D E, qui coule tout de mefme fur Y X en de-
meurant parallele a B C, D E pouffe E F, E F pouffe F G,
cellecy pouffe G H. & on en peut conceuoir vne infinité
d'autres', qui fe pouffent confequutiuement en mefme
façon, & dont les vnes facent toufiours les mefmes angles auec Y X, & les autres auec Y Z. Or pendant qu'on
ouure ainfi l'angle X Y Z, le point B defcrit la ligne A B,
qui eft vn cercle, & les autres poins D, F, H, ou fe font
les interfeçtions des autres reigles, defcriuent d'autres
lignes courbes A D, A F, A H, dont les dernieres font
par ordre plus côpofées que la premiere, & cellecy plus
que le cercle. mais ie ne voy pas ce qui peut empefcher,
qu'on ne concoiue auffy nettement, & auffy diftincte-
ment la defcription de cete premiere, que du cercle, ou
du

the ruler BC, fastened at right angles to XY at the point B, pushes toward Z the ruler CD which slides along YZ always at right angles. In like manner, CD pushes DE which slides along YX always parallel to BC; DE pushes EF; EF pushes FG; FG pushes GH, and so on. Thus we may imagine an infinity of rulers, each pushing another, half of them making equal angles with YX and the rest with YZ.

Now as the angle XYZ is increased the point B describes the curve AB, which is a circle; while the intersections of the other rulers, namely, the points D, F, H describe other curves, AD, AF, AH, of which the latter are more complex than the first and this more complex than the circle. Nevertheless I see no reason why the description of the first[73] cannot be conceived as clearly and distinctly as that of the circle, or at least as that of the conic sections; or why that of the second, third,[74] or any other that can be thus described, cannot be as clearly conceived of as the first: and therefore I see no reason why they should not be used in the same way in the solution of geometric problems.[75]

[73] That is, AD.

[74] That is, AF and AH.

[75] The equations of these curves may be obtained as follows: (1) Let $YA = YB = a$, $YC = x$, $CD = y$, $YD = z$; then $z : x = x : a$, whence $z = \dfrac{x^2}{a}$. Also $z^2 = x^2 + y^2$; therefore the equation of AD is $x^4 = a^2(x^2 + y^2)$. (2) Let $YA = YB = a$, $YE = x$, $EF = y$, $YF = z$. Then $z : x = x : YD$, whence $YD = \dfrac{x^2}{z}$. Also

$$x : YD = YD : YC, \text{ whence } YC = \frac{x^4}{z^2} \div x = \frac{x^3}{z^2}.$$

But $YD : YC = YC : a$, and therefore

$$\frac{ax^2}{z} = \left(\frac{x^3}{z^2}\right)^2, \text{ or } z = \sqrt[3]{\frac{x^4}{a}}.$$

Also, $z^2 = x^2 + y^2$. Thus we get, as the equation of AF,

$$\sqrt[3]{\frac{x^8}{a^2}} = x^2 + y^2, \text{ or } x^8 = a^2(x^2 + y^2)^3.$$

(3) In the same way, it can be shown that the equation of AH is

$$x^{12} = a^2(x^2 + y^2)^5.$$

See Rabuel, p. 107.

I could give here several other ways of tracing and conceiving a series of curved lines, each curve more complex than any preceding one,[76] but I think the best way to group together all such curves and then classify them in order, is by recognizing the fact that all points of those curves which we may call "geometric," that is, those which admit of precise and exact measurement, must bear a definite relation[77] to all points of a straight line, and that this relation must be expressed by means of a single equation.[78] If this equation contains no term of higher degree than the rectangle of two unknown quantities, or the square of one, the curve belongs to the first and simplest class,[79] which contains only the circle, the parabola, the hyperbola, and the ellipse; but when the equation contains one or more terms of the third or fourth degree[80] in one or both of the two unknown quantities[81] (for it requires two unknown quantities to express the relation between two points) the curve belongs to the second class; and if the equation contains a term of the fifth or sixth degree in either or both of the unknown quantities the curve belongs to the third class, and so on indefinitely.

[76] "Qui seroient de plus en plus composées par degrez à l'infini." The French quotations in the footnotes show a few variants in style in different editions.

[77] That is, a relation exactly known, as, for example, that between two straight lines in distinction to that between a straight line and a curve, unless the length of the curve is known.

[78] It will be recognized at once that this statement contains the fundamental concept of analytic geometry.

[79] "Du premier & plus simple genre," an expression not now recognized. As now understood, the order or degree of a plane curve is the greatest number of points in which it can be cut by any arbitrary line, while the class is the greatest number of tangents that can be drawn to it from any arbitrary point in the plane.

[80] Grouped together because an equation of the fourth degree can always be transformed into one of the third degree.

[81] Thus Descartes includes such terms as x^2y, x^2y^2, . . as well as x^3, y^4 . . . . .

du moins que des fections coniques; ny ce qui peut empefcher, qu'on ne concoiue la feconde, & la troifiefme, & toutes les autres, qu'on peut defcrire, auffy bien que la premiere; ny par confequent qu'on ne les recoiue toutes en mefme façon, pour feruir aux fpeculations de Geometrie.

Ie pourrois mettre icy plufieurs autres moyens pour tracer & conçeuoir des lignes courbes, qui feroient de plus en plus compofées par degrés a l infini. mais pour comprendre enfemble toutes celles, qui font en la nature, & les diftinguer par ordre en certains genres ; ie ne fçache rien de meilleur que de dire que tous les poins, de celles qu'on peut nommer Geometriques, c'eft a dire qui tombent fous quelque mefure précife & exacte, ont neceffairement quelque rapport a tous les poins d'vne ligne droite, qui peut eftre exprimé par quelque equation, en tous par vne mefme, Et que lorfque cete equation ne monte que iufques au rectangle de deux quantités indeterminées, oubien au quarré d'vne mefme, la ligne courbe eft du premier & plus fimple genre, dans lequel il ny a que le cercle, la parabole, l'hyperbole, & l'Ellipfe qui foient comprifes. mais que lorfque l'equation monte iufques a la trois ou quatriefme dimenfion des deux, ou de l'vne des deux quantités indeterminées, car il en faut deux pour expliquer icy le rapport d'vn point a vn autre, elle eft du fecond: & que lorfque l'equation monte iufques a la 5 ou fixiefme dimenfion, elle eft du troifiefme; & ainfi des autres a l'infini.

Comme fi ie veux fçauoir de quel genre eft la ligne E C, que i'imagine eftre defcrite par l'interfection de la

reigle

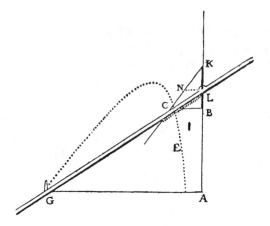

reigle G L, & du plan rectiligne C N K L, dont le costé
K N est indefiniement prolongé vers C , & qui estant
meu sur le plan de dessous en ligne droite, c'est a dire en
telle sorte que son diametre K L se trouue tousiours appliqué sur quelque endroit de la ligne B A prolongée de
part & d'autre, fait mouuoir circulairement cete reigle
G L autour du point G, a cause quelle luy est tellement
iointe quelle passe tousiours par le point L . Ie choisis
vne ligne droite, comme A B, pour rapporter a ses diuers
poins tous ceux de cete ligne courbe E C, & en cete ligne A B ie choisis vn point, comme A, pour commencer
par luy ce calcul. Ie dis que ie choisis & l'vn & l'autre, a
cause qu'il est libre de les prendre tels qu'on veult.   car
encore qu'il y ait beaucoup de choix pour rendre l'equation plus courte, & plus aysée; toutefois en quelle façon
qu'on les prene, on peut tousiours faire que la ligne pa-
roisse de mesme genre, ainsi qu'il est aysé a demonstrer.

Aprés

Suppose the curve EC to be described by the intersection of the ruler GL and the rectilinear plane figure CNKL, whose side KN is produced indefinitely in the direction of C, and which, being moved in the same plane in such a way that its side[82] KL always coincides with some part of the line BA (produced in both directions), imparts to the ruler GL a rotary motion about G (the ruler being hinged to the figure CNKL at L).[83] If I wish to find out to what class this curve belongs, I choose a straight line, as AB, to which to refer all its points, and in AB I choose a point A at which to begin the investigation.[84] I say "choose this and that," because we are free to choose what we will, for, while it is necessary to use care in the choice in order to make the equation as short and simple as possible, yet no matter what line I should take instead of AB the curve would always prove to be of the same class, a fact easily demonstrated.[85]

[82] "Diametre."

[83] The instrument thus consists of three parts, (1) a ruler AK of indefinite length, fixed in a plane; (2) a ruler GL, also of indefinite length, fastened to a pivot, G, in the same plane, but not on AK; and (3) a rectilinear figure BKC, the side KC being indefinitely long, to which the ruler GL is hinged at L, and which is made to slide along the ruler GL.

[84] That is, Descartes uses the point A as origin, and the line AB as axis of abscissas. He uses parallel ordinates, but does not draw the axis of ordinates.

[85] That is, the nature of a curve is not affected by a transformation of coördinates.

Then I take on the curve an arbitrary point, as C, at which we will suppose the instrument applied to describe the curve. Then I draw through C the line CB parallel to GA. Since CB and BA are unknown and indeterminate quantities, I shall call one of them y and the other x. To the relation between these quantities I must consider also the known quantities which determine the description of the curve, as GA, which I shall call a; KL, which I shall call b; and NL parallel to GA, which I shall call c. Then I say that as NL is to LK, or as c is to b, so CB, or y, is to BK, which is therefore equal to $\frac{b}{c}y$. Then BL is equal to $\frac{b}{c}y - b$, and AL is equal to $x + \frac{b}{c}y - b$. Moreover, as CB is to LB, that is, as y is to $\frac{b}{c}y - b$, so AG or a is to LA or $x + \frac{b}{c}y - b$. Multiplying the second by the third, we get $\frac{ab}{c}y - ab$ equal to

$$xy + \frac{b}{c}y^2 - by,$$

which is obtained by multiplying the first by the last. Therefore, the required equation is

$$y^2 = cy - \frac{cx}{b}y + ay - ac.$$

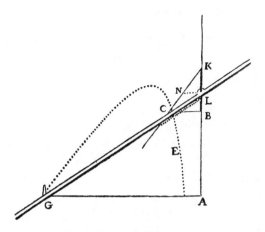

Aprés cela prenant vn point a diſcretion dans la courbe,
comme C, ſur lequel ie ſuppoſe que l'inſtrument qui ſert
a la deſcrire eſt appliqué, ie tire de ce point C·la ligne
C B parallele a G A, & pourceque C B & B A ſont deux
quantités indeterminées & inconnuës , ie les nomme
l'vne *y* & l'autre *x*. mais affin de trouuer le rapport de
l'vne à l'autre; ie conſidere auſſy les quantités connuës
qui determinent la deſcription de cete ligne courbe,
comme G A que ie nomme *a*, K L que ie nomme *b*, &
N L parallele à G A que ie nomme *c*. puis ie dis, comme
N L eſt à L K, ou *c* à *b*, ainſi C B, ou *y*, eſt à B K, qui eſt
par conſequent $\frac{b}{c} y$: & B L eſt $\frac{b}{c} y -- b$, & A L 'eſt *x* +
$\frac{b}{c} y -- b$. de plus comme C B eſt à L B, ou *y* à $\frac{b}{c} y -- b$, ainſi
a, ou G A, eſt á L A, ou *x* + $\frac{b}{c} y -- b$. de façon que mul-
S f   tipliant

53

tipliant la feconde par la troifiefme on produit $\frac{ab}{c}y -- ab$,

qui eft efgale à $xy + \frac{b}{c}yy -- by$ qui fe produit en multipliant la premiere par la derniere. & ainfi l'equation qu'il
falloit trouuer eft .

$$yy \infty cy -- \frac{cx}{b}y + ay -- ae.$$

de laquelle on connoift que la ligne E C eft du premier
genre , comme en effect elle n'eft autre qu'vne Hyperbole.

Que fi en l'inftrument qui fert a la defcrire on fait
qu'au lieu de la ligne droite C N K, ce foit cete Hyperbole, ou quelque autre ligne courbe du premier genre,
qui termine le plan C N K L; l'interfection de cete ligne
& de la reigle G L defcrira, au lieu de l'Hyperbole E C,
vne autre ligne courbe, qui fera du fecond genre. Comme fi C N K eft vn cercle, dont L foit le centre , on defcrira la premiere Conchoide des anciens ; & fi c'eft vne
Parabole dont le diametre foit K B, on defcrira la ligne
courbe, que i'ay tantoft dit eftre la premiere , & la plus
fimple pour la queftion de Pappus, lorfqu'il n'y a que cinq
lignes droites données par pofition. Mais fi au lieu d'vne
de ces lignes courbes du premier genre , c'en eft vne du
fecond, qui termine le plan C N K L, on en defcrira par
fon moyen vne du troifiefme, ou fi c'en eft vne du troifiefme, on en defcrira vne du quatriefme, & ainfi a l'infini.
comme il eft fort ayfé a connoiftre par le calcul. Et en
quelque autre façon, qu'on imagine la defcription d'vne
ligne courbe , pourvû qu'elle foit du nombre de celles
que ie nomme Geometriques , on pourra toufiours trouuer

From this equation we see that the curve EC belongs to the first class, it being, in fact, a hyperbola.[86]

If in the instrument used to describe the curve we substitute for the rectilinear figure CNK this hyperbola or some other curve of the first class lying in the plane CNKL, the intersection of this curve with the ruler GL will describe, instead of the hyperbola EC, another curve, which will be of the second class.

Thus, if CNK be a circle having its center at L, we shall describe the first conchoid of the ancients,[87] while if we use a parabola having KB as axis we shall describe the curve which, as I have already said, is the first and simplest of the curves required in the problem of Pappus, that is, the one which furnishes the solution when five lines are given in position.[88]

[86] Cf. Briot and Bouquet, *Elements of Analytical Geometry of Two Dimensions*, trans. by J. H. Boyd, New York, 1896, p. 143.
The two branches of the curve are determined by the position of the triangle CNKL with respect to the directrix AB. See Rabuel, p. 119.
Van Schooten, p. 171, gives the following construction and proof: Produce AG to D, making DG = EA. Since E is a point of the curve obtained when GL coincides with GA, L with A, and C with N, then EA = NL. Draw DF parallel to KC. Now let GCE be a hyperbola through E whose asymptotes are DF and FA. To prove that this hyperbola is the curve given by the instrument described above, produce BC to cut DF in I, and draw DH parallel to AF

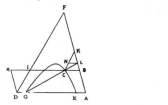

meeting BC in H. Then KL : LN = DH : HI. But DH = AB = x, so we may write $b : c = x$: HI, whence HI = $\frac{cx}{b}$, IB = $a + c - \frac{cx}{b}$, IC = $a + c - \frac{cx}{b} - y$.

But in any hyperbola IC . BC = DE . EA, whence we have $(a + c - \frac{cx}{b} - y)y = ac$,

or $y^2 = cy - \frac{cxy}{b} + ay - ac$. But this is the equation obtained above, which is

therefore the equation of a hyperbola whose asymptotes are AF and FD.
Van Schooten, p. 172, describes another similar instrument: Given a ruler AB pivoted at A, and another BD hinged to AB at B. Let AB rotate about A so that D moves along LK; then the curve generated by any point E of BE will be an ellipse whose semi-major axis is AB + BE and whose semi-minor axis is AB — BE.
[87] See notes 59 and 70.
[88] For a discussion of the elliptic, parabolic, and hyperbolic conchoids see Rabuel, pp. 123, 124.

If, instead of one of these curves of the first class, there be used a curve of the second class lying in the plane CNKL, a curve of the third class will be described; while if one of the third class be used, one of the fourth class will be obtained, and so on to infinity.[89] These statements are easily proved by actual calculation.

Thus, no matter how we conceive a curve to be described, provided it be one of those which I have called geometric, it is always possible to find in this manner an equation determining all its points. Now I shall place curves whose equations are of the fourth degree in the same class with those whose equations are of the third degree; and those whose equations are of the sixth degree[90] in the same class with those whose equations are of the fifth degree[91] and similarly for the rest. This classification is based upon the fact that there is a general rule for reducing to a cubic any equation of the fourth degree, and to an equation of the fifth degree[92] any equation of the sixth degree, so that the latter in each case need not be considered any more complex than the former.

It should be observed, however, with regard to the curves of any one class, that while many of them are equally complex so that they may be employed to determine the same points and construct the same problems, yet there are certain simpler ones whose usefulness is more limited. Thus, among the curves of the first class, besides the ellipse, the hyperbola, and the parabola, which are equally complex, there is also found the circle, which is evidently a simpler curve; while among those of the second class we find the common conchoid, which is described by means of the circle, and some others which, though less

[89] Rabuel (p. 125), illustrates this, substituting for the curve CNKL the semicubical parabola, and showing that the resulting equation is of the fifth degree, and therefore, according to Descartes, of the third class. Rabuel also gives (p. 119), a general method for finding the curve, no matter what figure is used for CNKL. Let $GA = a$, $KL = b$, $AB = x$, $CB = y$ and $KB = z$; then $LB = z - b$, and $AL = x + z - b$. Now $GA : AL = CB : BL$, or $a : x + z - b = y : z - b$, whence $z = \dfrac{xy - by + ab}{a - y}$.

This value of z is independent of the nature of the figure CNKL. But given any figure CNKL it is possible to obtain a second value for z from the nature of the curve. Equating these values of z we get the equation of the curve.

[90] "Celles dont l'équation monte au quarré de cube."
[91] "Celles dont elle ne monte qu'au sursolide."
[92] "Au sursolide."

uer vne equation pour déterminer tous ſes poins en cete
ſorte.

Au reſte ie mets les lignes courbes qui font monter
cete equation iuſques au quarré de quarré , au meſme
genre que celles qui ne la font monter que iuſques au
cube. & celles dont l'equation monte au quarré de cube, au meſme genre que celles dont elle ne monte qu'au
ſurſolide. & ainſi des autres. Dont la raiſon eſt, qu'il y a
reigle generale pour reduire au cube toutes les difficultés qui vont au quarré de quarré , & au ſurſolide toutes
celles qui vont au quarré de cube, de façon qu'on ne les
doit point eſtimer plus compoſées.

Mais il eſt a remarquer qu'entre les lignes de chaſque
genre, encore que la plus part ſoient eſgalement compoſées , en ſorte qu'elles peuuent ſeruir a déterminer les
meſmes poins, & conſtruire les meſmes probleſmes, il y
en a toutefois auſſy quelques vnes, qui ſont plus ſimples,
& qui n'ont pas tant d'eſtendue en leur puiſſance. comme entre celles du premier genre outre l'Ellipſe l'Hyperbole & la Parabole qui ſont eſgalement compoſées , le
cercle y eſt auſſy compris, qui manifeſtement eſt plus
ſimple. & entre celles du ſecond genre il y a la Conchoide vulgaire, qui a ſon origine du cercle; & il y en a encore quelques autres, qui bien qu'elles n'ayent pas tant
d'eſtendue que la plus part de celles du meſme genre,
ne peuuent toutefois eſtre miſes dans le premier.

Or aprés auoir ainſi reduit toutes les lignes courbes a
certains genres , il m'eſt ayſé de pourſuiure en la demonſtration de la reſponſe, que i'ay tantoſt faite a la queſtion de Pappus. Car premierement ayant fait voir cy
deſſus, Suite de l'explica-tion de la queſtion de Pappus miſe au liure pre-cedenr

S ſ 2

deſſus , que lorſqu'il n'y a que trois ou 4 lignes droites
données, l'equation qui ſert a determiner les poins cherchés, ne monte que iuſques au quarré; il eſt euident, que
la ligne courbe ou ſe trouuent ces poins, eſt neceſſairement quelqu vne de ceſles du premier genre: a cauſe que
cete meſme equation explique le rapport, qu'ont tous
les poins des lignes du premier genre a ceux d'vne ligne
droite. Et que lorſqu'il n'y a point plus de 8 lignes droites données , cete equation ne monte que iuſques au
quarré de quarré tout au plus , & que par conſequent la
ligne cherchée ne peut eſtre que du ſecond genre, ou au
deſſous. Et que lorſqu'il n'y a point plus de 12 lignes données , l'equation ne monte que iuſques au quarré de cube, & que par conſequent la ligne cherchée n'eſt que du
troiſieſme genre, ou au deſſous. & ainſi des autres. Et
meſme a cauſe que la poſition des lignes droites données
peut varier en toutes ſortes, & par conſequent faire chã-
ger tant les quantités connuës, que les ſignes + & -- de
l'equation, en toutes les façons imaginables ; il eſt euident qu'il n'y a aucune ligne courbe du premier genre,
qui ne ſoit vtile a cete queſtion, quand elle eſt propoſée
en 4 lignes droites; ny aucune du ſecond qui n y ſoit vtile, quand elle eſt propoſée en huit ; ny du troiſieſme,
quand elle eſt propoſée en douze: & ainſi des autres. En
ſorte qu'il n'y a pas vne ligne courbe qui tombe ſous le
calcul & puiſſe eſtre receüe en Geometrie , qui n'y ſoit
vtile pour quelque nombre de lignes.

Solution
de cete
queſtion
quandelle
n'eſt pro-
poſée
qu'en 3
ou 4 li-
gnes.

Mais il faut icy plus particulierement que ie determine, & donne la façon de trouuer la ligne cherchée ; qui
ſert en chaſque cas, lorſqu'il n y a que 3 ou 4 lignes droites

complicated[93] than many curves of the same class, cannot be placed in the first class.[94]

Having now made a general classification of curves, it is easy for me to demonstrate the solution which I have already given of the problem of Pappus. For, first, I have shown that when there are only three or four lines the equation which serves to determine the required points[95] is of the second degree. It follows that the curve containing these points must belong to the first class, since such an equation expresses the relation between all points of curves of Class I and all points of a fixed straight line. When there are not more than eight given lines the equation is at most a biquadratic, and therefore the resulting curve belongs to Class II or Class I. When there are not more than twelve given lines, the equation is of the sixth degree or lower, and therefore the required curve belongs to Class III or a lower class, and so on for other cases.

Now, since each of the given lines may have any conceivable position, and since any change in the position of a line produces a corresponding change in the values of the known quantities as well as in the signs + and — of the equation, it is clear that there is no curve of Class I that may not furnish a solution of this problem when it relates to four lines, and that there is no curve of Class II that may not furnish a solution when the problem relates to eight lines, none of Class III when it relates to twelve lines, etc. It follows that there is no geometric curve whose equation can be obtained that may not be used for some number of lines.[96]

It is now necessary to determine more particularly and to give the method of finding the curve required in each case, for only three or

[93] "Pas tant d'étenduë." Cf. Rabuel, p. 113. "Pas tant d'étendue en leur puissance."

[94] Various methods of tracing curves were used by writers of the seventeenth century. Among these there were not only the usual method of plotting a curve from its equation and that of using strings, pegs, etc., as in the popular construction of the ellipse, but also the method of using jointed rulers and that of using one curve from which to derive another, as for example the usual method of describing the cissoid. Cf. Rabuel, p. 138.

[95] That is, the equation of the required locus.

[96] "En sorte qu'il n'y a pas une ligne courbe qui tombe sous le calcul & puisse être receuë en Geometrie, qui n'y soit utile pour quelque nombre de lignes."

four given lines. This investigation will show that Class I contains only the circle and the three conic sections.

Consider again the four lines AB, AD, EF, and GH, given before, and let it be required to find the locus generated by a point C, such that, if four lines CB, CD, CF, and CH be drawn through it making given angles with the given lines, the product of CB and CF is equal to the product of CD and CH. This is equivalent to saying that if

$$CB = y,$$

$$CD = \frac{czy + bcx}{z^2},$$

$$CF = \frac{ezy + dek + dcx}{z^2},$$

and

$$CH = \frac{gzy + fgl - fgx}{z^2}.$$

then the equation is

$$y^2 = \frac{(cfglz - dckz^2)y - (dez^2 + cfgz - bcgz)xy + bcfglx - bcfgx^2}{ez^3 - cgz^2}.$$

res données; & on verra par mesme moyen que le premier genre des lignes courbes n'en contient aucunes autres, que les trois sections coniques, & le cercle.

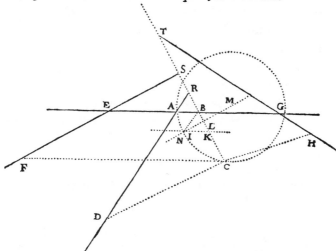

Reprenons les 4 lignes AB, AD, EF, & GH données cy deſſus, & qu'il faille trouuer vne autre ligne, en
laquelle il ſe rencontre vne infinité de poins tels que C,
duquel ayant tiré les 4 lignes CB, CD, CF, & CH, a
angles donnés, ſur les données, CB multipliée par CF,
produiſt une ſomme eſgale a CD, multipliée par CH.
c'eſt a dire ayant fait CB ∞ y, CD ∞ $\dfrac{czy \,\text{⅃ᴄ}\, bcx,}{zz}$

CF ∞ $\dfrac{ezy \,\text{⅃ᴄ}\, dek \,\text{⅃ᴄ}\, dex,}{zz}$ & CH ∞ $\dfrac{gzy \,\text{⅃ᴄ}\, fgl \,\text{--}\, fgx,}{zz}$ l'equatiõ eſt

$$yy \propto \left.\begin{array}{c} --dekzz \\ \text{⅃ᴄ}\, cfglz \end{array}\right\} \left.\begin{array}{c} --dezzx \\ --cfgzx \\ \text{⅃ᴄ}\, bcgzx \end{array}\right\} y \left.\begin{array}{c} \text{⅃ᴄ}\, bcfglx \\ --bcfgxx \end{array}\right\}$$

$$\overline{}$$

$$\dfrac{ezzz \,--\, cgzz.}{}$$

au

au moins en suppofant $e\,\chi$ plus grand que $e\,g$. car s'il eſtoit
moindre, il faudroit changer tous les ſignes $+$ & $--$. Et
ſi la quantité y ſe trouuoit nulle, ou moindre que rien en
cete equation, lorſqu'on a ſuppoſé le point C en l'angle
D A G, il faudroit le ſuppoſer auſſy en l'angle D A E, ou
E A R, ou R A G, en changeant les lignes $+$ & $--$ ſelon
qu'il ſeroit requis a cet effect. Et ſi en toutes ces 4 po-
ſitions la valeur d'y ſe trouuoit nulle, la queſtion ſeroit
impoſſible au cas propoſé. Mais ſuppoſons la icy eſtre
poſſible, & pour en abreger les termes, au lieu des quantités $\dfrac{cfglz--de\ell zz}{ez--cgzz}$ eſcriuons $2m$, & au lieu de

$\dfrac{dezz+cfgz--bcg\chi}{ez--cgz\chi}$ eſcriuons $\dfrac{2n}{z}$; & ainſi nous au-

rons

$$yy \infty 2my--\dfrac{2n}{z}xy\dfrac{+bcfglx--bcfgxx}{ez--cgzz},$$ dont la raci-

ne eſt

$$y \infty m--\dfrac{nx}{z}+\sqrt{mm--\dfrac{2mnx}{z}+\dfrac{nnxx}{z\chi}\dfrac{+bcfglx--bcfgxx}{ez--cgzz}}.$$

& derechef pour abreger, au lieu de

$$--\dfrac{2mn}{z}+\dfrac{bcfgl}{ez--cgzz}$$ eſcriuons o, & au lieu de $\dfrac{nn}{zz}\dfrac{--bcfg}{e--cgzz}$

eſcriuons $\dfrac{p}{m}$. car ces quantités eſtant toutes données,
nous les pouuons nommer comme il nous plaiſt. &
ainſi nous auons

$$y \infty m--\dfrac{n}{z}x+\sqrt{mm+ox--\dfrac{p}{m}xx},$$ qui doit eſtre la

longeur de la ligne B C, en laiſſant A B, ou x indeterminée.

It is here assumed that ez is greater than cg; otherwise the signs $+$ and $-$ must all be changed.[97] If y is zero or less than nothing in this equation,[98] the point C being supposed to lie within the angle DAG, then C must be supposed to lie within one of the angles DAE, EAR, or RAG, and the signs must be changed to produce this result. If for each of these four positions y is equal to zero, then the problem admits of no solution in the case proposed.

Let us suppose the solution possible, and to shorten the work let us write $2m$ instead of $\dfrac{cflgz - dekz^2}{ez^3 - cgz^2}$, and $\dfrac{2n}{z}$ instead of $\dfrac{dez^2 + cfgz - bcga}{ez^3 - cgz^2}$. Then we have

$$y^2 = 2my - \frac{2n}{z}xy + \frac{bcfglx - bcfgx^2}{ez^3 - cgz^2},$$

of which the root[99] is

$$y = m - \frac{nx}{z} + \sqrt{m^2 - \frac{2mnx}{z} + \frac{n^2x^2}{z^2} + \frac{bcfglx - bcfgx^2}{ez^3 - cgz^2}}.$$

Again, for the sake of brevity, put $-\dfrac{2mn}{z} + \dfrac{bcfgl}{ez^3 - cgz^2}$ equal to o, and $\dfrac{n^2}{z^2} - \dfrac{bcfg}{ez^3 - cgz^2}$ equal to $\dfrac{p}{m}$; for these quantities being given, we can represent them in any way we please.[100] Then we have

$$y = m - \frac{n}{z}x + \sqrt{m^2 + ox + \frac{p}{m}x^2}.$$

This must give the length of the line BC, leaving AB or x undeter-

[97] When ez is greater than cg, then $ez^3 - cgz^2$ is positive and its square root is therefore real.

[98] Descartes uses "moindre que rien" for "negative."

[99] Descartes mentions here only one root; of course the other root would furnish a second locus.

[100] In a letter to Mersenne (Cousin, Vol. VII, p. 157), Descartes says: "In regard to the problem of Pappus, I have given only the construction and demonstration without putting in all the analysis; . . . in other words, I have given the construction as architects build structures, giving the specifications and leaving the actual manual labor to carpenters and masons."

mined. Since the problem relates to only three or four lines, it is obvious that we shall always have such terms, although some of them may vanish and the signs may all vary.[101]

After this, I make KI equal and parallel to BA, and cutting off on BC a segment BK equal to m (since the expression for BC contains $+ m$; if this were $- m$, I should have drawn IK on the other side of AB,[102] while if m were zero, I would not have drawn IK at all). Then I draw IL so that IK : KL $= z : n$; that is, so that if IK is equal to x, KL is equal to $\frac{n}{z}x$. In the same way I know the ratio of KL to IL, which I may call $n : a$, so that if KL is equal to $\frac{n}{z}x$, IL is equal to $\frac{a}{z}x$. I take the point K between L and C, since the equation contains $-\frac{n}{z}x$; if this were $+\frac{n}{z}x$, I should take L between K and C;[103] while if $\frac{n}{z}x$ were equal to zero, I should not draw IL.

This being done, there remains the expression

$$LC = \sqrt{m^2 + ox + \frac{p}{m}x^2},$$

from which to construct LC. It is clear that if this were zero the point

[101] Having obtained the value of BC algebraically, Descartes now proceeds to construct the length BC geometrically, term by term. He considers BC equal to BK + KL + LC, which is equal to BK — LK + LC which in turn is equal to

$$m - \frac{n}{z}x + \sqrt{m^2 + ox + \frac{p}{m}x^2}.$$

[102] That is, take I on CB produced.

[103] That is, on KB produced. C is not yet determined.

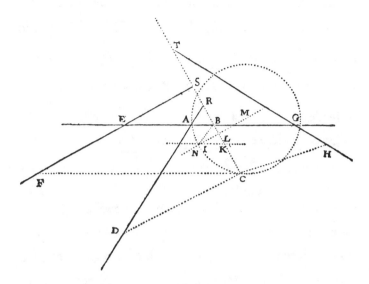

minée. Et il eſt euident que la queſtion n'eſtant propoſée qu'en trois ou quatre lignes, on peut touſiours
auoir de tels termes. excepté que quelques vns d'eux
peuuent eſtre nuls, & que les ſignes + & -- peuuent di-
uerſement eſtre changés.

Aprés cela ie fais K I eſgale & parallele a B A, en ſorte
qu'elle couppe de B C la partie B K eſgale à *m*, à cauſe
qu'il y a icy + *m*; & ie l'aurois adiouſtée en tirant cete
ligne I K de l'autre coſté, s'il y auoit eu -- *m*; & ie ne l'aurois point du tout tirée, ſi la quantité *m* euſt eſté nullé.
Puis ie tire auſſy I L, en ſorte que la ligne I K eſt a K L,
comme Z eſt a *n*. c'eſt a dire que I K eſtant *x*, K L eſt
$\frac{n}{z}x$. Et par meſme moyen ie connois auſſy la proportion

qui

qui eſt entre K L, & I L, que ie poſe comme entre n & a:
ſi bien que K L eſtant $\frac{n}{z}x$, I L eſt $\frac{a}{z}x$; Et ie fais que le
point K ſoit entre L & C, a cauſe qu'il y a icy $--\frac{n}{z}x$;
au lieu que i'aurois mis L entre K & C, ſi i'euſſe eu $+-\frac{n}{z}x$;
& ie n'euſſe point tiré cete ligne I L, ſi $\frac{n}{z}x$ euſt eſté nulle.

Or cela fait, il ne me reſte plus pour la ligne L C, que
ces termes, L C $\infty\sqrt{mm + ox -- \frac{p}{m}xx}$. d'où ie voy
que s'ils eſtoient nuls, ce point C ſe trouueroit en la ligne droite I L; & que s'ils eſtoient tels que la racine s'en
puſt tirer, c'eſt a dire que mm & $\frac{p}{m}xx$ eſtant marqués
d'vn meſme ſigne $+-$ ou $--$, oo fuſt eſgal à $4pm$, ou bien
que les termes mm & ox, ou ox & $\frac{p}{m}xx$ fuſſent nuls, ce
point C ſe trouueroit en vne autre ligne droite qui ne ſeroit pas plus malayſée a trouuer qu' I L. Mais lorſque
cela n'eſt pas, ce point C eſt touſiours en l'une des trois
ſections coniques, ou en vn cercle, dont l'vn des diametres eſt en la ligne I L, & la ligne L C eſt l'vne de celles qui s'appliquent par ordre à ce diametre; ou au contraire L C eſt parallele au diametre, auquel celle qui eſt
en la ligne I L eſt appliquée par ordre. A ſçavoir ſi le terme $\frac{p}{m}xx$, eſt nul cete ſection conique eſt vne Parabole;
& s'il eſt marqué du ſigne $+$, c'eſt vne Hyperbole; &
enfin s'il eſt marqué du ſigne $--$ c'eſt vne Ellipſe. Excepté
ſeulement ſi la quantité aam eſt eſgale à pzz & que l'angle I L C ſoit droit : auquel cas on à vn cercle au lieu
d'vne

C would lie on the straight line IL;[104] that if it were a perfect square, that is if m^2 and $\frac{p}{m} x^2$ were both $+$[105] and o^2 was equal to $4pm$, or if m^2 and ox, or ox and $\frac{p}{m} x^2$, were zero, then the point C would lie on another straight line, whose position could be determined as easily as that of IL.[106]

If none of these exceptional cases occur,[107] the point C always lies on one of the three conic sections, or on a circle having its diameter in the line IL and having LC a line applied in order to this diameter,[108] or, on the other hand, having LC parallel to a diameter and IL applied in order.

In particular, if the term $\frac{p}{m} x^2$ is zero, the conic section is a parabola; if it is preceded by a plus sign, it is a hyperbola; and, finally, if it is preceded by a minus sign, it is an ellipse.[109] An exception occurs when

[104] The equation of IL is $y = m - \frac{n}{z} x$.

[105] There is considerable diversity in the treatment of this sentence in different editions. The Latin edition of 1683 has "Hoc est, ut, mm & $\frac{p}{m} xx$ signo $+$ notalis." The French edition, Paris, 1705, has "C'est à dire que mm et $\frac{p}{m} xx$ étant marquez d'un même signe $+$ ou $-$." Rabuel gives "C'est a dire que mm and $\frac{p}{m} xx$ étant marquez d'un même signe $+$." He adds the following note: "Il y a dans les Editions Françoises de Leyde, 1637, et de Paris, 1705, 'un meme signe $+$ ou $-$', ce qui est une faute d'impression." The French edition, Paris, 1886, has "Etant marqués d'un meme signe $+$ ou $-$."

[106] Note the difficulty in generalization experienced even by Descartes. Cf. Briot and Bouquet, p. 72.

[107] "Mais lorsque cela n'est pas." In each case the equation giving the value of y is linear in x and y, and therefore represents a straight line. If the quantity under the radical sign and $\frac{n}{z} x$ are both zero, the line is parallel to AB. If the quantity under the radical sign and m are both zero, C lies in AL.

[108] "An ordinate." The equivalent of "ordination application" was used in the 16th century translation of Apollonius. Hutton's Mathematical Dictionary, 1796, gives "applicate." "Ordinate applicate," was also used.

[109] Cf. Briot and Bouquet, p. 143.

a^2m is equal to pz^2 and the angle ILC is a right angle,[110] in which case we get a circle instead of an ellipse.[111]

If the conic section is a parabola, its latus rectum is equal to $\dfrac{oz}{a}$ and its axis always lies along the line IL.[112] To find its vertex, N, make IN equal to $\dfrac{am^2}{oz}$, so that the point I lies between L and N if m^2 is positive and ox is positive; and L lies between I and N if m^2 is positive and ox negative; and N lies between I and L if m^2 is negative and ox positive. It is impossible that m^2 should be negative when the terms' are arranged as above. Finally, if m^2 is equal to zero, the points N and I must coincide. It is thus easy to determine this parabola, according to the first problem of the first book of Apollonius[113].

If, however, the required locus is a circle, an ellipse, or a hyperbola,[114] the point M, the center of the figure, must first be found. This

[110] Rabuel (p. 167) adds "If $a^2m = pz^2$ or if $m = p$ the hyperbola is equilateral."

[111] In this case the triangle ILK is a right triangle, whence $\overline{IK}^2 = \overline{LK}^2 + \overline{IC}^2$; but by hypothesis IL : IK : KL $= a : z : n$; then $a^2 + n^2 = z^2$. Now the equation of the curve is

$$y = m - \frac{n}{z} + x\sqrt{m^2 + oz - \frac{p}{m}x^2},$$

and therefore the term in x^2 is

$$\left(\frac{n^2}{z^2} + \frac{p}{m}\right)x^2;$$

and if $a^2m = pz^2$, then $\dfrac{p}{m} = \dfrac{a^2}{z^2}$, and this term in x^2 becomes $\dfrac{a^2 + n^2}{z^2}x^2 = x^2$.

Therefore, the coefficients of x^2 and y^2 are unity and the locus is a circle.

[112] This may be seen as follows: From the figure, and by the nature of the parabola $\overline{LC}^2 = LN.p$ and LN $=$ IL$+$IN. Let IN $= \phi$; then since IL $= \dfrac{a}{z}x$, we have LN $= \dfrac{a}{z}x + \phi$ and LC $= y - m + \dfrac{n}{z}x$; whence $(y - m + \dfrac{n}{z}x)^2 = (\dfrac{a}{z}x + \phi)p$. But $(y - m + \dfrac{n}{z}x)^2 = m^2 + ox$ from the equation of the parabola; therefore $\dfrac{a}{z}xp + \phi p = m^2 + ox$. Equating coefficients, we have $\dfrac{a}{z}p = o$; $p = \dfrac{oz}{a}$; $\phi p = m^2$; $\phi\dfrac{oz}{a} = m^2$; $\phi = \dfrac{am^2}{oz}$.

[113] *Apollonii Pergaei Quae Graece exstant* edidit I. L. Heiberg, Leipzig, 1891. Vol. I, p. 159, Liber I, Prop. LII. Hereafter referred to as Apollonius. This may be freely translated as follows: To describe in a plane a parabola, having given the parameter, the vertex, and the angle between an ordinate and the corresponding abscissa.

[114] Central conics are thus grouped together by Descartes, the circle being treated as a special form of the ellipse, but being mentioned separately in all cases.

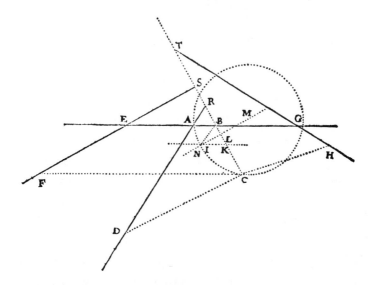

d'vne Ellipſe. Que ſi cete ſection eſt vne Parabole, ſon coſté droit eſt eſgal à $\frac{o\,\zeta}{a}$, & ſon diametre eſt touſiours en la ligne I L. & pour trouuer le point N, qui en eſt le ſommet, il faut faire I N eſgale à $\frac{a\,m\,m}{o\,z}$; & que le point I ſoit entre L & N, ſi les termes ſont $+ m\,m + o\,x$; oubien que le point L ſoit entre I & N, s'ils ſont $+ m\,m -- o\,x$; oubien il faudroit qu'N fuſt entré I & L, s'il y auoit $-- m\,m + o\,x$. Mais il ne peut iamais y auoir $-- m\,m$, en la façon que les termes ont icy eſté poſés. Et enfin le point N ſeroit le meſme que le point I ſi la quantité $m\,m$ eſtoit nulle. Au moyen dequoy il eſt ayſé de trouuer cete Parabole par le 1er. Probleſme du 1er. liure d'Apollonius.

T t

Que

Que si la ligne demãdée est vn cercle, ou vne ellipse, ou vne Hyperbole, il faut premierement chercher le point M, qui en est le centre, & qui est tousiours en la ligne droite I L, ou on le trouue en prenant $\frac{aom}{2p\zeta}$ pour I M. en forte que si la quantité o est nulle, ce centre est iustement au point I. Et si la ligne cherchée est vn cercle, ou vne Ellipse; on doit prendre lé point M du mesmé costé que le point L, au respect du point I, lorsqu'on a $+ox$; & lorsqu'on à $--ox$, on le doit prendre de l'autre. Mais tout au contraire en l'Hyperbole, si on a $--ox$, ce centre M doit estre vers L; & si on a $+ox$, il doit estre de l'autre costé. Aprés cela le costé droit de la figure doit estre

$$\sqrt{\frac{ooz\,z}{aa}+\frac{4mpzz,}{aa}}$$

lorsqu'on a $+mm$, & que la ligne cherchée est vn cercle, ou vne Ellipse; oubien lorsqu'on à $--mm$, & que c'est vne Hyperbole. & il doit estre

$$\sqrt{\frac{oozz}{aa}-\frac{4mpzz,}{aa}}$$

si la ligne cherchée estant vn cercle, ou vne Ellipse, ou à $--mm$; oubien si estant vne Hyperbole & la quantité oo estant plus grande que $4mp$, on à $+mm$. Que si la quantité mm est nulle, ce costé droit est $\frac{o\zeta}{a}$, & si o x est nulle, il est $\sqrt{\frac{4mpz\zeta}{aa}}$. Puis pour le costé traversant, il faut trouuer vne ligne, qui soit a ce costé droit, cõme aam est à $p\zeta\zeta$, à sçauoir si ce costé droit est

$$\sqrt{\frac{oozz}{aa}+\frac{4mpzz,}{aa}}\qquad\text{le trauersant est}\qquad\sqrt{\frac{aaoomm}{ppzz}+\frac{4aam}{pzz}}.$$

Et en tous ces cas le diametre de la section est en la ligne I M, & L C est l'vne de celles qui luy est appliquée par ordre. Sibienque faisant M N esgale a la moitié du costé

trauer-

will always lie on the line IL and may be found by taking IM equal to $\dfrac{aom}{2pz}$.[115] If o is equal to zero M coincides with I. If the required locus is a circle or an ellipse, M and L must lie on the same side of I when the term ox is positive and on opposite sides when ox is negative. On the other hand, in the case of the hyperbola, M and L lie on the same side of I when ox is negative and on opposite sides when ox is positive.

The latus rectum of the figure must be

$$\sqrt{\frac{o^2z^2}{a^2} + \frac{4mpz^2}{a^2}}$$

if m^2 is positive and the locus is a circle or an ellipse, or if m^2 is negative and the locus is a hyperbola. It must be

$$\sqrt{\frac{o^2z^2}{a^2} - \frac{4mpz^2}{a^2}}$$

if the required locus is a circle or an ellipse and m^2 is negative, or if it is an hyperbola and o^2 is greater than $4mp$, m^2 being positive.

But if m^2 is equal to zero, the latus rectum is $\dfrac{oz}{a}$; and if oz is equal to zero[116], it is

$$\sqrt{\frac{4mpz^2}{a^2}}.$$

For the corresponding diameter a line must be found which bears the ratio $\dfrac{a^2m}{pz^2}$ to the latus rectum; that is, if the latus rectum is

$$\sqrt{\frac{o^2z^2}{a^2} + \frac{4mpz^2}{a^2}}$$

the diameter is

$$\sqrt{\frac{a^2o^2m^2}{p^2z^2} + \frac{4a^2m^3}{pz^2}}.$$

In every case, the diameter of the section lies along IM, and LC is one of its lines applied in order.[117] It is thus evident that, by making MN equal to half the diameter and taking N and L on the same side of M,

[115] Cf. Briot and Bouquet, p. 156.
[116] Some editions give, incorrectly, ox for oz.
[117] See note 108.

71

the point N will be the vertex of this diameter.[118] It is then a simple matter to determine the curve, according to the second and third problems of the first book of Apollonius.[119]

When the locus is a hyperbola[120] and m^2 is positive, if o^2 is equal to zero or less than $4pm$ we must draw the line MOP from the center M parallel to LC, and draw CP parallel to LM, and take MO equal to

$$\sqrt{m^2 - \frac{o^2 m}{4p}};$$

while if ox is equal to zero, MO must be taken equal to m. Then considering O as the vertex of this hyperbola, the diameter being OP and the line applied in order being CP, its latus rectum is

$$\sqrt{\frac{4a^4 m^4}{p^2 z^4} - \frac{a^4 o^2 m^3}{p^3 z^4}}$$

and its diameter[121] is

$$\sqrt{4m^2 - \frac{o^2 m}{p}}.$$

[118] If the equation contains $-m^2$ and $+nx$, then n^2 must be greater than $4mp$, otherwise the problem is impossible.

[119] Cf. Apollonius, Vol. I, p. 173, Lib. I, Prop. LV: To describe a hyperbola, given the axis, the vertex, the parameter, and the angle between the axes. Also see Prop. LVI: To describe an ellipse, etc.

[120] Cf. Letters of Descartes, Cousin, Vol. VIII, p. 142.

[121] "Côté traversant."

trauerſant & le prenant du meſme coſté du point M,
qu'eſt le point L, on a le point N pour le ſommet de ce
diametre. en ſuite dequoy il eſt ayſé de trouuer la ſection
par le ſecond & 3 prob. du 1ᵉʳ. liu. d'Apollonius.

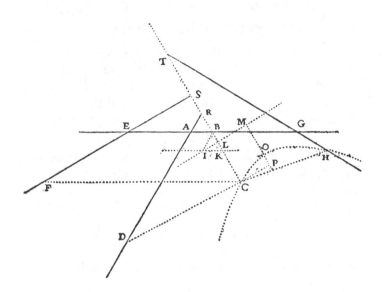

Mais quand cete ſection eſtant vne Hyperbole, on à
+ m m; & que la quantité o o eſt nulle ou plus petite que
4 p m, on doit tirer du centre M la ligne M O P parallele a
L C, & C P parallele à L M: & faire M O eſgale a

$$\sqrt{m\,m - \frac{o\,o\,m}{4\,p}}\,;$$ oubien la faire eſgale à m ſi la quantité o x
eſt nulle. Puis conſiderer le point O, côme le ſommet
de cete Hyperbole; dont le diametre eſt O P, & C P la

ligne qui luy est appliquée par ordre, & son costé droit est

$$\sqrt{\frac{4 \, d^4 m^4}{p p z^4} - \frac{a^4 o o m^3}{p^3 z^4}} \quad \text{\& son costé trauersãt est} \quad \sqrt{4 m m - \frac{o o m}{p}}$$

Excepté quand $o x$ est nulle. car alors le costé droit est $\frac{2 \, a a m m}{p z z}$, & le trauersant est $2 m$. & ainsi il est aysé de la trouuer par le 3 prob. du 1er. liu. d'Apollonius.

Demon stration de tout ce qui vient d'estre expliqué. Et les demonstrations de tout cecy sont euidentes. car composant vn espace des quantités que iay assignées pour le costé droit, & le trauersant, & pour le segment du diametre N L, ou O P, suiuãt la teneur de l'11, du 12, & du 13 theoresmes du 1er. liure d'Apollonius, on trouuera tous les mesmes termes dont est composé le quarré de la ligne C P, ou C L, qui est appliquée par ordre a ce diametre. Comme en cet exemple ostant I M , qui est $\frac{a o m}{2 p z}$, de N M, qui est $\frac{a m}{2 p z} \sqrt{o o + 4 m p}$, iay I N, a laquelle aioustant I L, qui est $\frac{a}{z} x$, iay N L , qui est $\frac{a}{z} x - \frac{a o m}{2 p z}$ $+ \frac{a m}{2 p z} \sqrt{o o + 4 m p}$, & cecy estant multiplié par $\frac{z}{a} \sqrt{o o + 4 m p}$, qui est le costé droit de la figure, il vient

$$x \sqrt{o o + 4 m p} - \frac{o m}{2 p} \sqrt{o o + 4 m p} + \frac{m o o}{2 p} + 2 m m$$

pour le rectangle. duquel il faut oster vn espace qui soit au quarré de N L comme le costé droit est au trauersant. & ce quarré de N L est $\frac{a a}{z z} x x - \frac{a a o m}{p z z} x$

$$+ \frac{a a m}{p z z} x \sqrt{o o + 4 m p} + \frac{a a o o m m}{2 p p z z} + \frac{a a m}{p z z}$$
$$- \frac{a a o m m}{2 p p z z}.$$

An exception must be made when ox is equal to zero, in which case the latus rectum is $\dfrac{2a^2m^2}{pz^2}$ and the diameter is $2m$. From these data the curve can be determined in accordance with the third problem of the first book of Apollonius.[122]

The demonstrations of the above statements are all very simple, for, forming the product[123] of the quantities given above as latus rectum, diameter, and segment of the diameter NL or OP, by the methods of Theorems 11, 12, and 13 of the first book of Apollonius, the result will contain exactly the terms which express the square of the line CP or CL, which is an ordinate of this diameter.

In this case take IM or $\dfrac{aom}{2pz}$ from NM or from its equal

$$\frac{am}{2pz}\ \sqrt{o^2+4mp}.$$

To the remainder IN add IL or $\dfrac{a}{z}x$, and we have

$$NL=\frac{a}{z}x-\frac{aom}{2pz}+\frac{am}{2pz}\ \sqrt{o^2+4mp}.$$

Multiplying this by

$$\frac{z}{a}\ \sqrt{o^2+4mp},$$

the latus rectum of the curve, we get

$$x\ \sqrt{o^2+4mp}-\frac{om}{2p}\ \sqrt{o^2+4mp}+\frac{mo^2}{2p}+2m^2$$

for the rectangle, from which is to be subtracted a rectangle which is to the square of NL as the latus rectum is to the diameter. The square of NL is

$$\frac{a^2}{z^2}x^2-\frac{a^2om}{pz^2}x+\frac{a^2m}{pz^2}x\ \sqrt{o^2+4mp}+\frac{a^2o^2m^2}{2p^2z^2}+\frac{a^2m^3}{pz^2}-\frac{a^2om^2}{2p^2z^2}\ \sqrt{o^2+4mp}.$$

[122] See note 113.
[123] "Composant un espace."

Divide this by a^2m and multiply the quotient by pz^2, since these terms express the ratio between the diameter and the latus rectum. The result is

$$\frac{p}{m}x^2 - ox + x\sqrt{o^2 + 4mp} + \frac{o^2m}{2p} - \frac{om}{2p}\sqrt{o^2 + 4mp} + m^2.$$

This quantity being subtracted from the rectangle previously obtained, we get

$$\overline{CL}^2 = m^2 + ox - \frac{p}{m}x^2.$$

It follows that CL is an ordinate of an ellipse or circle applied to NL, the segment of the axis.

Suppose all the given quantities expressed numerically, as $EA=3$, $AG=5$, $AB=BR$, $BS = \frac{1}{2}BE$, $GB=BT$, $CD=\frac{3}{2}CR$, $CF=2CS$, $CH = \frac{2}{3}CT$, the angle $ABR=60°$; and let $CB.CF=CD.CH$. All these quanties must be known if the problem is to be entirely determined. Now let $AB=x$, and $CB=y$. By the method given above we shall obtain

$$y^2 = 2y - xy + 5x - x^2;$$

$$y = 1 - \frac{1}{2}x + \sqrt{1 + 4x - \frac{3}{4}x^2};$$

whence BK must be equal to 1, and KL must be equal to one-half KI; and since the angle $IKL =$ angle $ABR = 60°$ and angle KIL (which is one-half angle KIB or one-half angle IKL) is $30°$, the angle ILK is a right angle. Since $IK=AB=x$, $KL=\frac{1}{2}x$, $IL=x\sqrt{\frac{3}{4}}$, and the quantity represented by z above is 1, we have $a=\sqrt{\frac{3}{4}}$, $m=1$, $o=4$, $p=\frac{3}{4}$, whence $IM=\sqrt{\frac{16}{3}}$, $NM=\sqrt{\frac{19}{3}}$; and since a^2m (which is $\frac{3}{4}$) is equal to pz^2, and

$$-\;-\frac{a\,a\,o\,m\,m}{2\,p\,p\,z\,z}\;\sqrt{\;00+4\,m\,p\;}$$ qu'il faut diuifer par *a a m* &

multiplier par *p z z*, a caufe que ces termes expliquent la
proportion qui eft entre le cofté trauerfant & le droit, &

il vient $\frac{p}{m}\,x\,x\,-\,o\,x\,+\,x\,\sqrt{\;00+4\,m\,p\;}\,+\,\frac{o\,o\,m}{2\,p}$

$-\,\frac{o\,m}{2\,p}\,\sqrt{\;00+4\,m\,p\;}\,+\,mm.$ ce qu'il faut ofter du rectangle

precedent, & on trouue $mm+ox-\frac{p}{m}\,xx$ pour le quarré de C L, qui par confequent eft vne ligne appliquée
par ordre dans vne Ellipfe, ou dans vn cercle, au fegment
du diametre N L.

Et fi on vent expliquer toutes les quantités données
par nombres, en faifant par exemple E A ∞ 3, A G ∞ 5,
A B ∞ B R, B S ∞ $\frac{1}{2}$ B E, G B ∞ B T, C D ∞ $\frac{2}{2}$ C R, C F
∞ 2 C S, C H ∞ $\frac{2}{3}$ C T, & que l angle A B R foit de 60
degrés; & enfin que le rectangle des deux C B, & C F,
foit efgal au rectangle des deux autres C D & C H; car il
faut auoir toutes ces chofes affin que la queftion foit entierement determinée. & auec cela fuppofant A B ∞ *x*;
& C B ∞ *y*, on trouue par la façon cy deffus expliquée
$y\,y$ ∞ $2\,y$ $-$ $x\,y$ $+$ $5\,x$ $-$ $x\,x$ & y ∞ 1 $-$ $\frac{1}{2}x$ $+$
$\sqrt{\;1+4\,x-\frac{2}{4}\,x\,x}$: fi bien que B K doit eftre 1, & K L
doit eftre la moitié de K I, & pourceque l'angle I K L
ou A B R eft de 60 degrés, & K I L qui eft la moitié de
K I B ou I K L, de 30, I L K eft droit. Et pourceque I K
ou A B eft nommé *x*, K L eft $\frac{1}{2}x$, & I L eft $x\sqrt{\frac{2}{4}}$, & la
quantité qui eftoit tantoft nommée *z* eft 1, celle qui
eftoit *a* eft $\sqrt{\frac{2}{4}}$, celle qui eftoit *m* eft 1, celle qui eftoit *o*
eft 4, & celle qui eftoit *p* eft $\frac{2}{4}$, de façon qu'on à $\sqrt{\;1\frac{6}{3}}$

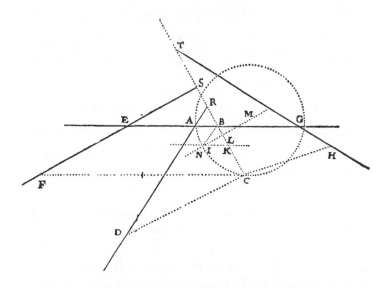

pour I M, & $\sqrt{}$ $^{1}_{3}\!^{2}$ pour N M, & pourceque *aam* qui
eſt$\frac{3}{4}$ eſt icy eſgal à *pzz* & que l'angle I L C eſt droit, on
trouue que la ligne courbe N C eſt vn cercle. Et on
peut facilement examiner tous les autres cas en meſme
ſorte.

Quels
ſont les
lieux
plans, &
ſolides: &
la facon
de les
trouuer.
Au reſte a cauſe que les equations, qui ne montent
que iuſques au quarré, ſont toutes compriſes en ce que ie
viens d'expliquer; non ſeulement le probleſme des anciens en 3 & 4 lignes eſt icy entierement acheué; mais
auſſy tout ce qui appartient à ce qu'ils nommoient la
compoſition des lieux ſolides; & par conſēquent auſſy a
celle des lieux plans, a cauſe qu'ils ſont compris dans les
ſolides. Car ces lieux ne ſont autre choſe, ſinon que lors
qu'il eſt queſtion de trouuer quelque point auquel il
manque

the angle ILC is a right angle, it follows that the curve NC is a circle. A similar treatment of any of the other cases offers no difficulty.

Since all equations of degree not higher than the second are included in the discussion just given, not only is the problem of the ancients relating to three or four lines completely solved, but also the whole problem of what they called the composition of solid loci, and consequently that of plane loci, since they are included under solid loci.[124] For the solution of any one of these problems of loci is nothing more than the finding of a point for whose complete determination one con-

---

[124] Since plane loci are degenerate cases of solid loci. The case in which neither x^2 nor y^2 but only xy occurs, and the case in which a constant term occurs, are omitted by Descartes. The various kinds of solid loci represented by the equation $y = \pm m \pm \dfrac{n}{z} x \pm \dfrac{n^2}{x} \pm \sqrt{\pm m^2 \pm ox \pm \dfrac{p}{m} x}$ may be summarized as follows:
(1) If all the terms of the right member are zero except $\dfrac{n^2}{x}$, the equation represents an hyperbola referred to its asymptotes. (2) If $\dfrac{n^2}{x}$ is not present, there are several cases, as follows: (a) If the quantity under the radical sign is zero or a perfect square, the equation represents a straight line; (b) If this quantity is not a perfect square and if $\dfrac{p}{m} x^2 = 0$, the equation represents a parabola; (c) If it is not a perfect square and if $\dfrac{p}{m} x^2$ is negative, the equation represents a circle or an ellipse; (d) If $\dfrac{p}{m} x^2$ is positive, the equation represents a hyperbola. Rabuel, p. 248.

dition is wanting, the other conditions being such that (as in this example) all the points of a single line will satisfy them. If the line is straight or circular, it is said to be a plane locus; but if it is a parabola. a hyperbola, or an ellipse, it is called a solid locus. In every such case an equation can be obtained containing two unknown quantities and entirely analogous to those found above. If the curve upon which the required point lies is of higher degree than the conic sections, it may be called in the same way a supersolid locus,[125] and so on for other cases. If two conditions for the determination of the point are lacking, the locus of the point is a surface, which may be plane, spherical, or more complex. The ancients attempted nothing beyond the composition of solid loci, and it would appear that the sole aim of Apollonius in his treatise on the conic sections was the solution of problems of solid loci.

I have shown, further, that what I have termed the first class of curves contains no others besides the circle, the parabola, the hyperbola, and the ellipse. This is what I undertook to prove.

[125] "Un lieu sursolide."

manque vne condition pour eftre entierement determiné, ainfi qu'il arriue en cete exemple, tous les poins d'vne
mefme ligne peuuent eftre pris pour celuy qui eft demandé. Et fi cete ligne eft droite, ou circulaire , on la
nomme vn lieu plan. Mais fi c'eft vne parabole , ou vne
hyperbole, ou vne ellipfe, on la nomme vn lieu folide. Et
toutefois & quantes que cela eft, on peut venir a vne E-
quation qui contient deux quantités inconnuës , & eft
pareille a quelqu'vne de celles que ie viens de refoudre.
Que fi la ligne qui determine ainfi le point cherché , eft
d'vn degré plus compofée que les fections coniques, on
la peut nommer, en mefme façon , vn lieu furfolide, &
ainfi des autres. Et s'il manque deux conditions a la determination de ce point, le lieu ou il fe trouue eft vne fuperficie, laquelle peut eftre tout de mefme ou plate, ou
fpherique , ou plus compofée. Mais le plus haut but
qu'ayent eu les anciens en cete matiere a efté de parue-
nir a la compofition des lieux folides : Et il femble que
tout ce qu'Apollonius a efcrit des fections coniques n'a
efté qu'à deffein de la chercher.

De plus on voit icy que ceque iay pris pour le premier
genre des lignes courbes, n'en peut comprendre aucunes
autres que le cercle, la parabole, l'hyperbole, & l'ellipfe.
qui eft tout ce que i'auois entrepris de prouuer.

Que fi la queftion des anciens eft propofée en cinq lignes, qui foient toutes paralleles ; il eft euident que le
point cherché fera toufiours en vne ligne droite · Mais fi
elle eft propofée en cinq lignes, dont il y en ait quatre
qui foient paralleles, & que la cinquiefme les couppe a
angles droits, & mefme que toutes les lignes tirées du
point

Quelle eft la premie-
re & la plus fim-
ple de toutes les
lignes courbes qui fer-
uent en la queftion des anciens quand el-
le eft pro-
pofée en cinq li-
gnes.

point cherché les rencontrent auſſy a angles droits, &
enfin que le parallelepipede compoſé de trois des lignes
ainſi tirées ſur trois de celles qui ſont paralleles, ſoit eſgal
au parallelepipede compoſé des deux lignes tirées l'vne
ſur la quatrieſme de celles qui ſont paralleles & l'autre
ſur celle qui les couppe a angles droits, & d'vne troiſieſme ligne donnée. ce qui eſt ce ſemble le plus ſimple cas qu'on puiſſe imaginer aprés le precedent ; le
point cherché ſera en la ligne courbe, qui eſt deſcrite
par le mouuement d'vne parabole en la façon cy deſſus
expliquée.

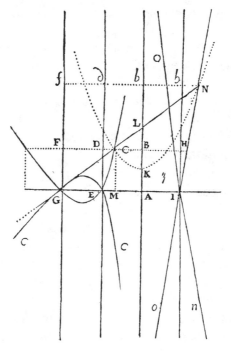

Soient

If the problem of the ancients be proposed concerning five lines, all parallel, the required point will evidently always lie on a straight line. Suppose it be proposed concerning five lines with the following conditions:

(1) Four of these lines parallel and the fifth perpendicular to each of the others,

(2) The lines drawn from the required point to meet the given lines at right angles;

(3) The parallelepiped[126] composed of the three lines drawn to meet three of the parallel lines must be equal to that composed of three lines, namely, the one drawn to meet the fourth parallel, the one drawn to meet the perpendicular, and a certain given line.

This is, with the exception of the preceding one, the simplest possible case. The point required will lie on a curve generated by the motion of a parabola in the following way:

[126] That is, the product of the numerical measures of these lines.

Let the required lines be AB, IH, ED, GF, and GA, and let it be required to find the point C, such that if CB, CF, CD, CH, and CM be drawn perpendicular respectively to the·given lines, the parallelepiped of the three lines CF, CD, and CH shall be equal to that of the other two, CB and CM, and a third line AI. Let CB$=y$, CM$=x$, AI or AE or GE$=a$; whence if C lies between AB and DE, we have CF$=2a-y$, CD$=a-y$, and CH$=y+a$. Multiplying these three together we get $y^3-2ay^2-a^2y+2a^3$ equal to the product of the other three, namely to axy.

I shall consider next the curve CEG, which I imagine to be described by the intersection of the parabola CKN (which is made to move so that its axis KL always lies along the straight line AB) with the ruler GL (which rotates about the point G in such a way that it constantly lies in the plane of the parabola and passes through the point L). I take KL equal to a and let the principal parameter, that is, the parameter corresponding to the axis of the given parabola, be also equal to a, and let GA$=2a$, CB or MA$=y$, CM or AB$=x$. Since the triangles GMC and CBL are similar, GM (or $2a-y$) is to MC (or x) as CB (or y) is to BL, which is therefore equal to $\dfrac{xy}{2a-y}$. Since KL is a, BK is $a-\dfrac{xy}{2a-y}$ or $\dfrac{2a^2-ay-xy}{2a-y}$. Finally, since this same BK is a segment of the axis of the parabola, BK is to BC (its ordinate) as BC is to a (the latus rectum), whence we get $y^3-2ay^2-a^2y+2a^3=axy$, and therefore C is the required point.

Soient par exemple les lignes cherchées A B, I H, E D, G F, & G A. & qu'on demande le point C, en sorte que tirant C B, C F, C D, C H, & C M a angles droits sur les données, le parallelepipede des trois C F, C D, & C H soit esgal a celuy des 2 autres C B, & C M, & d'vne troisiesme qui soit A I. Ie pose C B ∞ y. C M ∞ x. A I, ou A E, ou G E ∞ a, de façon que le point C estant entre les lignes A B, & D E, iay C F ∞ 2 a -- y, C D ∞ a -- y. & C H ∞ y + a. & multipliant ces trois l'vne par l'autre, iay y^3 -- 2 a y y -- a a y + 2 a^3 esgal au produit des trois autres qui est a x y. Aprés cela ie considere la ligne courbe C E G, que i'imagine estre descrite par l'intersection, de la Parabole C K N, qu'on fait mouuoir en telle sorte que son diametre K L est tousiours sur la ligne droite A B, & de la reigle G L qui tourne cependant autour du point G en telle sorte quelle passe tousiours dans le plan de cete Parabole par le point L. Et ie fais K L ∞ a, & le costé droit principal, c'est a dire celuy qui se rapporte a l'aissieu de cete parabole, aussy esgal à a, & G A ∞ 2 a, & C B ou M A ∞ y, & C M ou A B ∞ x. Puis a cause des triangles semblables G M C & C B L, G M qui est 2 a - y, est à M C qui est x, comme C B qui est y, est à B L qui est par consequent $\frac{x y}{2 a -- y}$. Et pourceque L K est a, B K est a $\frac{-- x y}{2 a - y}$, oubien $\frac{2 a a -- a y -- x y}{2 a -- y}$. Et enfin pourceque ce mesme B K estant vn segment du diametre de la Parabole, est à B C qui luy est appliquée par ordre, comme cellecy est au costé droit qui est a, le calcul monstre que y -- 2 a y y -- a a y + 2 a, est esgal à a x y. & par conse-

V v   quent

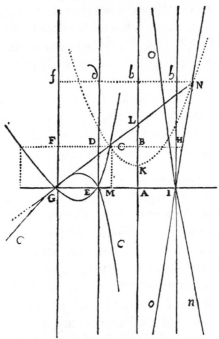

quent que le point C eſt celuy qui eſtoit demandé. Et il
peut eſtre pris en tel endroit de la ligne C E G qu'on veuille choiſir, ou auſſy en ſon adiointe *c* E G *c* qui ſe deſcrit en meſme façon, excepté que le ſommet de la Parabole eſt tourné vers l'autre coſté, ou enfin en leurs con-
trepoſées N I *o, n* I O, qui ſont deſcrites par l'interſection
que fait la ligne G L en l'autre coſté de la Parabole
K N.

    Or encore que les paralleles données A B, I H, E D,
& G F ne fuſſent point eſgalement diſtantes, & que G A
ne les couppaſt point a angles droits, ny auſſy les lignes
                                       tirées

The point C can be taken on any part of the curve CEG or of its adjunct cEGc, which is described in the same way as the former, except that the vertex of the parabola is turned in the opposite direction; or it may lie on their counterparts[127] NIo and nIO, which are generated by the intersection of the line GL with the other branch of the parabola KN.

Again, suppose that the given parallel lines AB, IH, ED, and GF are not equally distant from one another and are not perpendicular to GA, and that the lines through C are oblique to the given lines. In this case the point C will not always lie on a curve of just the same nature. This may even occur when no two of the given lines are parallel.

[127] "En leurs contreposées."

Next, suppose that we have four parallel lines, and a fifth line cutting them, such that the parallelepiped of three lines drawn through the point C (one to the cutting line and two to two of the parallel lines) is equal to the parallelepiped of two lines drawn through C to meet the other two parallels respectively and another given line. In this case the required point lies on a curve of different nature,[128] namely, a curve such that, all the ordinates to its axis being equal to the ordinates of a conic section, the segments of the axis between the vertex and the ordinates[129] bear the same ratio to a certain given line that this line bears to the segments of the axis of the conic section having equal ordinates.[130]

I cannot say that this curve is less simple than the preceding; indeed, I have always thought the former should be considered first, since its description and the determination of its equation are somewhat easier.

I shall not stop to consider in detail the curves corresponding to the other cases, for I have not undertaken to give a complete discussion of the subject; and having explained the method of determining an infinite number of points lying on any curve, I think I have furnished a way to describe them.

It is worthy of note that there is a great difference between this method[131] in which the curve is traced by finding several points upon

[128] The general equation of this curve is $axy - xy^2 + 2a^2x = a^2y - ay^2$. Rabuel, p. 270.

[129] That is, the abscissas of points on the curve.

[130] The thought, expressed in modern phraseology, is as follows: The curve is of such nature that the abscissa of any point on it is a third proportional to the abscissa of a point on a conic section whose ordinate is the same as that of the given point, and a given line. Cf. Rabuel, pp. 270, et seq.

[131] That is, the method of analytic geometry.

tirées du point C vers elles, ce point Ċ ne laiſſeroit pas
de ſe trouuer touſiours en vne ligne courbe, qui ſeroit
de cete meſme nature. Et il s'y peut auſſy trouuer quelquefois, encore qu'aucune des lignes données ne ſoient
paralleles. Mais ſi lorſqu'il y en a 4 ainſi paralleles, & vne
cinquieſme qui les trauerſe: & que le parallelepipede de
trois des lignes tirées du point cherché, l'vne ſur cete
cinquieſme, & les 2 autres ſur 2 de celles qui ſont paralleles; ſoit eſgal a celuy, des deux tirées ſur les deux autres paralleles , & d'vne autre ligne donnée. Ce point
cherché eſt en vne ligne courbe d'vne autre nature, a
ſçauoir en vne qui eſt telle, que toutes les lignes droites
appliquées par ordre a ſon diametre eſtant eſgales a celles d'vne ſeċtion conique, les ſegmens de ce diametre,
qui ſont entre le ſommet & ces lignes , ont meſme proportion a vne certaine ligne donnée, que cete ligne donnée a aux ſegmens du diametre de la ſeċtion conique,
auſquels les pareilles lignes ſont appliquées par ordre. Et
ie ne ſçaurois veritablement dire que cete ligne ſoit
moins ſimple que la precedente, laquelle iay creu toutefois deuoir prendre pour la premiere, a cauſe que la deſcription, & le calcul en ſont en quelque façon plus
faciles.

　　Pour les lignes qui ſeruent aux autres cas, ie ne m'are-
ſteray point a les diſtinguer par eſpeces. car ie n'ay pas
entrepris de dire tout ; & ayant expliqué la façon de
trouuer vne infinité de poins par ou elles paſſent, ie penſe
auoir aſſés donné le moyen de les deſcrire.

　　Meſme il eſt a propos de remarquer, qu'il y a grande
difference entre cete façon de trouuer pluſieurs poins

Quelles
font les
lignes
courbes
qu'on de-
fcrit en
trouuant
plufieurs
de leurs
poins, qui
peuuent
eftre re-
ceues en
Geome-
trie.

pour tracer vne ligne courbe, & celle dont on fe fert pour
la fpirale, & fes femblables. car par cete derniere on ne
trouue pas indifferēment tous les poins de la ligne qu'on
cherche, mais feulement ceux qui peuuent eftre déter-
minés par quelque mefure plus fimple, que celle qui eft
requife pour la compofer, & ainfi a proprement parler
on ne trouue pas vn de fes poins. c'eft a dire pas vn de
ceux qui luy font tellement propres, qu'ils ne puiffent
eftre trouués que par elle: Au lieu qu'il ny a aucun point
dans les lignes quiferuent a la queftion propofée, qui ne
fe puiffe rencontrer entre ceux qui fe determinent par la
façon tañtoft expliquée. Et pourceque cete façon de
tracer une ligne courbe, en trouuant indifferēment plufieurs de fes poins, ne s'eftend qu'a celles qui peuuent
auffy eftre defcrites par vn mouuement regulier & continu, on ne la doit pas entierement reietter de la Geometrie.

Quelles
font auffy
cèlles
qu'on de-
fcrit auec
vne chor-
de, qui
peuuent
y eftre
receues.

Et on n'en doit pas reietter non plus, celle ou on fe
fert d'vn fil, ou d'vne chorde repliée, pour determiner
l'egalité ou la difference de deux ou plufieurs lignes
droites qui peuuent eftre tirées de chafque point de la
courbe qu'on cherche, a certains autres poins , ou fur
certaines autres lignes a certains angles. ainfi que nous
auons fait en la Dioptrique pour expliquer l'Ellipfe &
l'Hyperbole. car encore qu'on n'y puiffe reçeuoir aucunes lignes qui femblent a des chordes, c'eft a dire qui
deuienent tantoft droites & tantoft courbes, a caufe que
la proportion, qui eft entre les droites & les courbes,
n'eftant pas connuë, & mefme ie croy ne le pouuant eftre
par les hommes, on ne pourroit rien conclure de là qui
fuft

it, and that used for the spiral and similar curves.[132] In the latter not any point of the required curve may be found at pleasure, but only such points as can be determined by a process simpler than that required for the composition of the curve. Therefore, strictly speaking, we do not find any one of its points, that is, not any one of those which are so peculiarly points of this curve that they cannot be found except by means of it. On the other hand, there is no point on these curves which supplies a solution for the proposed problem that cannot be determined by the method I have given.

But the fact that this method of tracing a curve by determining a number of its points taken at random applies only to curves that can be generated by a regular and continuous motion does not justify its exclusion from geometry. Nor should we reject the method[133] in which a string or loop of thread is used to determine the equality or difference of two or more straight lines drawn from each point of the required curve to certain other points,[134] or making fixed angles with certain other lines. We have used this method in "La Dioptrique" [135] in the discussion of the ellipse and the hyperbola.

On the other hand, geometry should not include lines that are like strings, in that they are sometimes straight and sometimes curved, since the ratios between straight and curved lines are not known, and I believe cannot be discovered by human minds,[136] and therefore no conclusion based upon such ratios can be accepted as rigorous and exact.

[132] That is, transcendental curves, called by Descartes "mechanical" curves.

[133] Cf. the familiar "mechanical descriptions" of the conic sections.

[134] As for example, the foci, in the description of the ellipse.

[135] This work was published at Leyden in 1637, together with Descartes's *Discours de la Methode*.

[136] This is of course concerned with the problem of the rectification of curves. See Cantor, Vol. II (1), pp. 794 and 807, and especially p. 778. This statement, "ne pouvant être par les hommes" is a very noteworthy one, coming as it does from a philosopher like Descartes. On the philosophical question involved, consult such writers as Bertrand Russell.

Nevertheless, since strings can be used in these constructions only to determine lines whose lengths are known, they need not be wholly excluded.

When the relation between all points of a curve and all points of a straight line is known,[137] in the way I have already explained, it is easy to find the relation between the points of the curve and all other given points and lines; and from these relations to find its diameters, axes, center and other lines[138] or points which have especial significance for this curve, and thence to conceive various ways of describing the curve, and to choose the easiest.

By this method alone it is then possible to find out all that can be determined about the magnitude of their areas,[139] and there is no need for further explanation from me.

[137] Expressed by means of the equation of the curve.

[138] For example, the equations of tangents, normals, etc.

[139] For the history of the quadrature of curves, consult Cantor, Vol. II (1), pp. 758, et seq., Smith, *History,* Vol. II, p. 302.

fuſt exact & aſſuré. Toutefois a cauſe qu'on ne ſe ſert
de chordes en ces conſtructions, que pour déterminer
des lignes droites, dont on connoiſt parfaitement la longeur, cela ne doit point faire qu'on les reiette.

Or de cela ſeul qu'on ſçait le rapport, qu'ont tous les
poins d'vne ligne courbe a tous ceux d'vne ligne droite,
en la façon que iay expliquée; il eſt ayſé de trouuer auſſy
le rapport qu'ils ont a tous les autres poins, & lignes données: & en ſuite de connoiſtre les diametres, les aiſſieux,
les centres, & autres lignes, ou poins, a qui chaſque ligne courbe aura quelque rapport plus particulier, ou
plus ſimple, qu'aux autres: & ainſi d'imaginer diuers
moyens pour les deſcrire, & d'en choiſir les plus faciles.
Et meſme on peut auſſy par cela ſeul trouuer quaſi tout
ce qui peut eſtre determiné touchant la grandeur de l'e-
ſpace quelles comprenent, ſans qu'il ſoit beſoin que i'en
donne plus d'ouuerture. Et enfin pour cequi eſt de toutes les autres proprietés qu'on peut attribuer aux lignes
courbes, elles ne dependent que de la grandeur des angles qu'elles font auec quelques autres lignes. Mais lorſqu'on peut tirer des lignes droites qui les couppent a angles droits, aux poins ou elles ſont rencontrées par cel-
lésauec qui elles font les angles qu'on veut meſurer, ou,
ceque ie prens icy pour le meſme, qui couppent leurs
contingentes; la grandeur de ces angles n'eſt pas plus
malayſée a trouuer, que s'ils eſtoient compris entre deux
lignes droites. C'eſt pourquoy ie croyray auoir mis icy
tout ce qui eſt requis pour les elemens des lignes courbes, lorſque i'auray generalement donné la façon de tirer des lignes droites, qui tombent a angles droits ſur

Que pour-
trouuer
toutes les-
proprie-
tes des li-
gnes
courbes,
il ſuffiſt
de ſcauoir.
le rapport
qu'ont
tous leurs
poins a
ceux des
lignes
droites,
& la façon
de tirer
d'autres
lignes
qui les
couppent
en tous
ces poins
a angles
droits.

V v. 3   tels

tels de leurs poins qu'on voudra choifir. Et i'ofe dire
que c'eſt cecy le probléſme le plus vtile, & le plus general non feulement que ie ſçache, mais meſme que i'aye
iamais defiré de ſçauoir en Geometrie.

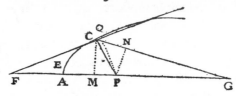

Facon
generale
pour
trouuer
des lignes
droites,
qui coup-
pent les
courbes
données,
ou leurs
contin-
gentes, a
angles
droits.

Soit C E
la ligne courbe,
& qu'il faille ti-
rer vne ligne
droite par le
point C, qui fa-
ce auec elle des angles droits. Ie ſuppoſe la choſe defia
faite, & que la ligne cherchée eſt C P, laquelle ie prolonge iuſques au point P, ou elle rencontre la ligne droite G A, que ie ſuppoſe eſtre celle aux poins de laquelle
on rapporte tous ceux de la ligne C E : en ſorte que faiſant M A ou C B ∞ y, & C M, ou B A ∞ x, iay quelque
equation, qui explique le rapport, qui eſt entre x & y.
Puis ie fais P C ∞ s, & P A ∞ v, ou P M ∞ v -- y, & a
cauſe du triangle rectangle P M C iay ss, qui eſt le quar-
ré de la baze eſgal à $xx + vv -- 2vy + yy$, qui ſont
les quarrés des deux coſtés. c'eſt a dire iay x ∞
$\sqrt{ss -- vv + 2vy -- yy}$, oubien y ∞ v + $\sqrt{ss -- xx}$, &
par le moyen de cete equation, i'oſte de l'autre equa-
tion qui m'explique le rapport qu'ont tous les poins de la
courbe C E a ceux de la droite G A, l'vne des deux quantités indeterminées x ou y. ce qui eſt ayſé a faire en
mettant partout $\sqrt{ss -- vv + 2vy -- yy}$ au lieu d'x, &
le quarré de cete ſomme au lieu d'x x, & ſon cube au lieu
d'x^3, & ainſi des autres, ſi c'eſt x que ie veuille oſter ; oubien

Finally, all other properties of curves depend only on the angles which these curves make with other lines. But the angle formed by two intersecting curves can be as easily measured as the angle between two straight lines, provided that a straight line can be drawn making right angles with one of these curves at its point of intersection with the other.[140] This is my reason for believing that I shall have given here a sufficient introduction to the study of curves when I have given a general method of drawing a straight line making right angles with a curve at an arbitrarily chosen point upon it. And I dare say that this is not only the most useful and most general problem in geometry that I know, but even that I have ever desired to know.

Let CE be the given curve, and let it be required to draw through C a straight line making right angles with CE. Suppose the problem solved, and let the required line be CP. Produce CP to meet the straight line GA, to whose points the points of CE are to be related.[141] Then, let $MA = CB = y$; and $CM = BA = x$. An equation must be found expressing the relation between x and y.[142] I let $PC = s$, $PA = v$, whence $PM = v - y$. Since PMC is a right triangle, we see that s^2, the square of the hypotenuse, is equal to $x^2 + v^2 - 2vy + y^2$, the sum of the squares of the two sides. That is to say, $x = \sqrt{s^2 - v^2 + 2vy - y^2}$ or $y = v + \sqrt{s^2 - x^2}$. By means of these last two equations, I can eliminate one of the two quantities x and y from the equation expressing the relation between the points of the curve CE and those of the straight line GA. If x is to be eliminated, this may easily be done by replacing x wherever it occurs by $\sqrt{s^2 - v^2 + 2vy - y^2}$, x^2 by the square of this expression, x^3 by its cube, etc., while if y is to be eliminated, y must be replaced by $v + \sqrt{s^2 - x^2}$, and $y^2, y^3, \ldots$ by the square of this expres-

[140] That is, the angle between two curves is defined as the angle between the normals to the curve at the point of intersection.

[141] That is, the line GA is taken as one of the coördinate axes.

[142] This will be the equation of the curve. See also the figure on page 97.

sion, its cube, and so on. The result will be an equation in only one unknown quantity, x or y.

For example, if CE is an ellipse, MA the segment of its axis of which CM is an ordinate, r its latus rectum, and q its transverse axis,[143] then by Theorem 13, Book I, of Apollonius,[144] we have

$x^2 = ry - \dfrac{r}{q}y^2$. Eliminating x^2 the resulting equation is

$$s^2 - v^2 + 2vy - y^2 = ry - \frac{r}{q}y^2, \quad \text{or} \quad y^2 + \frac{qry - 2qvy + qv^2 - qs^2}{q - r} = 0.$$

In this case it is better to consider the whole as constituting a single expression than as consisting of two equal parts.[145]

If CE be the curve generated by the motion of a parabola (see pages 47, et seq.) already discussed, and if we represent GA by b, KL by c, and the parameter of the axis KL of the parabola by d, the equation

[143] "Le traversant."

[144] Apollonius, p. 49: "Si conus per axem plano secatur autem alio quoque plano, quod cum utroque latere trianguli per axem posita concurrit, sed neque basi coni parallelum ducitur neque e contrario et si planum, in quo est basis coni, planumque secans concurrunt in recta perpendiculari aut ad basim trianguli per axem positi aut ad eam productam quælibet recta, quæ a sectione coni communi sectioni planorum parallela ducitur ad diametrum sectiones sumpta quadrata æqualis erit spatio adplicato rectæ cuidam, ad quam diametrus sectionis rationem habet, quam habet quadratum rectæ a vertice coni diametro sectionis parallelæ ductæ usque ad basim trianguli ad rectangulum comprehensum rectis ab ea ad latera trianguli abscissis, latitudinem rectam ab ea e diametro ad verticem sectionis abscissam et figura deficiens simili similiterque posita rectangulo a diametro parametroque comprehenso; vocetur autem talis sectio ellipsis." Cf. *Apollonius of Perga*, edited by Sir T. L. Heath, Cambridge, 1896, p. 11.

[145] That is, to transpose all the terms to the left member.

bien si c'est y, en mettant en son lieu $x + \sqrt{ss - xx}$, & le quarré, ou le cube, &c. de cete somme, au lieu d'yy, ou y^3 &c. De façon qu'il reste tousiours aprés cela vne equation, en laquelle il ny a plus qu'vne seule quantité indeterminée, x, ou y.

Comme si C E est vne Ellipse, & que M A soit le segment de son diametre, auquel C M soit appliquée par ordre, & qui ait r pour son costé droit, & q pour le trauersant, on à par le 13 th. du 1 liu. d'Apollonius.

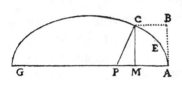

$xx \infty ry - \frac{r}{q}yy$, d'on oftant xx, il reste $ss -$

$- vv + 2vy - yy \infty ry - \frac{r}{q}yy.$

oubien,

$$yy \frac{+\ qry - 2qvy\ + \ qvv - qss}{q - r}\ \text{esgal a rien.}$$ car il est mieux en cet endroit de considerer ainsi ensemble toute la somme, que d'en faire vne partie esgale a l'autre.

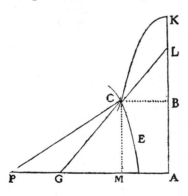

Tout de mesme si C E est la ligne courbe descrite par le mouuement d'vne Parabole en la façon cy dessus expliquée, & qu'on ait posé b pour G A, c pour K L, & d pour le costé droit du diametre K L en la parabole: l'equatiõ qui explique le rapport qui

qui eſt entre x & y, eſt $y^3 -- byy -- cdy + bcd + dxy \infty 0$.

d'où oſtant x, on a $y -- byy -- cdy + bcd + dy$
$\sqrt{ss -- vv + 2vy -- yy}$. & remetranr en ordre ces
termes par le moyen de la multiplication, il vient

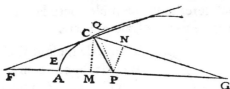

Et ainſi des autres.

Meſme encore que les poins de la ligne courbe ne ſe
rapportaſſent pas en la façon que iay ditte a ceux d'vne
ligne droite, mais en toute autre qu'on ſçauroit imaginer, on ne laiſſe pas de pouuoir touſiours auoir vne telle
equation. Comme ſi Ç E eſt vne ligne, qui ait tel rapport aux trois poins F, G, & A, que les lignes droites tirées de chaſcun de ſes poins comme C, iuſques au point
F, ſurpaſſent la ligne F A d'vne quantité, qui ait certaine
proportiõ don-
née a vne autre
quantité dont
GA ſurpaſſe les
lignes tirées
des meſmes
poins iuſques à G. Faiſons G A ∞ b, A F ∞ c, & prenant
à diſcretion le point C dans la courbe, que la quantité
dont C F ſurpaſſe F A, ſoit à celle dont G A ſurpaſſe
G C, commè d à e, en ſorte que ſi cete quantité qui eſt
indeterminée ſe nomme z, FC eſt $c + z$, & GC eſt $b -- \frac{e}{d}z$.
Puis poſant M A ∞ y, G M eſt $b -- y$, & F M eſt $c + y$, &
a cauſe du triangle rectangle C M G, oſtant le quarré
de

expressing the relation between x and y is $y^3 - by^2 - cdy + bcd + dxy = 0$. Eliminating x, we have

$$y^3 - by^2 - cdy + bcd + dy \sqrt{s^2 - v^2 + 2vy - y^2} = 0.$$

Arranging the terms according to the powers of y by squaring,[146] this becomes

$$y^6 - 2by^5 + (b^2 - 2cd + d^2)y^4 + (4bcd - 2d^2v)y^3$$
$$+ (c^2d^2 - d^2s^2 + d^2v^2 - 2b^2cd)y^2 - 2bc^2d^2y + b^2c^2d^2 = 0,$$

and so for the other cases. If the points of the curve are not related to those of a straight line in the way explained, but are related in some other way,[147] such an equation can always be found.

Let CE be a curve which is so related to the points F, G, and A, that a straight line drawn from any point on it, as C, to F exceeds the line FA by a quantity which bears a given ratio to the excess of GA over the line drawn from the point C to G.[148] Let $GA = b$, $AF = c$, and taking an arbitrary point C on the curve let the quantity by which CF exceeds FA be to the quantity by which GA exceeds GC as d is to e. Then if we let z represent the undetermined quantity, $FC = c + z$ and $GC = b - \frac{e}{d}z$. Let $MA = y$, $GM = b - y$, and $FM = c + y$. Since CMG is a right triangle, taking the square of GM from the square of GC we have

---

[146] "En remettant en ordre ces termes par moyen de la multiplication."

[147] "Mais en toute autre qu'on saurait imaginer."

[148] That is the ratio of CF — FA to GA — CG is a constant.

left the square of CM, or $\dfrac{e^2}{d^2}z^2 - \dfrac{2be}{d}z + 2by - y^2$. Again, taking the square of FM from the square of FC we have the square of CM expressed in another way, namely: $z^2 + 2cz - 2cy - y^2$. These two expressions being equal they will yield the value of y or MA, which is

$$\frac{d^2z^2 + 2cd^2z - e^2z^2 + 2bdez}{2bd^2 + 2cd^2}.$$

Substituting this value for y in the expression for the square of CM, we have

$$\overline{\text{CM}}^2 = \frac{bd^2z^2 + ce^2z^2 + 2bcd^2z - 2bcdez}{bd^2 + cd^2} - y^2.$$

If now we suppose the line PC to meet the curve at right angles at C, and let PC$=s$ and PA$=v$ as before, PM is equal to $v-y$; and since PCM is a right triangle, we have $s^2 - v^2 + 2vy - y^2$ for the square of CM. Substituting for y its value, and equating the values of the square of CM, we have

$$z^2 + \frac{2bcd^2z - 2bcdez - 2cd^2vz - 2bdevz - bd^2s^2 + bd^2v^2 - cd^2s^2 + cd^2v^2}{bd^2 + ce^2 + e^2v - d^2v} = 0$$

for the required equation.

Such an equation having been found[149] it is to be used, not to determine x, y, or z, which are known, since the point C is given, but to find v or s, which determine the required point P. With this in view, observe that if the point P fulfills the required conditions, the circle about P as center and passing through the point C will touch but not cut the curve CE; but if this point P be ever so little nearer to or farther from A than it should be, this circle must cut the curve not only

[149] Three such equations have been found by Descartes, namely those for the ellipse, the parabolic conchoid, and the curve just described.

de G M du quarré de G C, on a le quarré de C M, qui eſt
$\frac{ee}{dd}zz - \frac{2be}{d}z + 2by - yy$. puis oſtant le quarré de F M
du quarré de F C, on a encore le quarré de C M en d'au-
tres termes, a ſçauoir $zz + 2cz - 2cy - yy$, & ces termes eſtant eſgaux aux precedens, ils font connoiſtre y,
ou M A, qui eſt $\dfrac{ddzz + 2cddz - eezz + 2bdez}{2bdd + 2cdd}$ & ſubſtituant cete ſomme au lieu d'y dans le quarré de C M, on trouue
qu'il s'exprime en ces termes.

$$\frac{bddzz + ceezz + 2bcddz - 2bcdez}{bdd + cdd} - yy.$$

Puis ſuppoſant que la ligne droite P C rencontre la
courbe a angles droits au point C, & faiſant P C ∞ s, &
P A ∞ v comme deuant, P M eſt $v - y$; & a cauſe du
triangle rectangle P C M, on à $ss - vv + 2vy - yy$ pour
le quarré de C M, ou derechef ayant au lieu d'y ſubſtitué
la ſomme qui luy eſt eſgale, il vient

$$zz \frac{+ 2bcddz - 2bcdez - 2cddvz - 2bdevz - bddss + bddvv -}{bdd + cee \quad ee v - dd v}$$
$$- cddss + cddvv. \propto o \text{ pour l'equation que nous cherchions.}$$

Or aprés qu'on à trouué vne telle equation , au lieu
de s'en ſeruir pour connoiſtre les quantités x, ou y, ou z,
qui font deſia données, puiſque le point C eſt donné, on
la doit employer a trouuer v, ou s, qui determinent le
point P, qui eſt demandé. Et a cet effect il faut conſiderer, que ſi ce point P eſt tel qu'on le deſire, le cercle dont
il ſera le centre, & qui paſſera par le point C, y touchera
la ligne courbe C E, ſans la coupper: mais que ſi ce point
P, eſt tant ſoit peu plus proche, ou plus eſloigné du point
X x   A, qu'il

A, qu'il ne doit, ce cercle couppera la courbe , non feulement au point C, mais auſſy neceſſairement en quelque autre. Puis il faut auſſy conſiderer, que lorſque ce
cercle couppe la ligne courbe C E, l'equation par laquelle on cherche la quantité x, ou y, ou quelque autre ſemblable, en ſuppoſant P A & P C eſtre connuës, contient
neceſſairement deux racines, qui ſont ineſgales. Car par
exemple ſi ce cercle couppe la courbe aux poins C & E,
ayant tiré E Q parallele a C M, les noms des quantités
indeterminées x & y, conuiendront auſſy bien aux lignes
E Q, & Q A, qu'a C M, & M A; puis P E eſt eſgale a
P C, a cauſe du cercle, ſi bien que cherchant les lignes

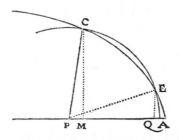

E Q & Q A, par P E &
P A qu'on ſuppoſe com-
me données , on aura la
meſme equation, que ſi
on cherchoit C M &
M A par P C, P A. d'où
il ſuit euideı̃ment, que la
valeur d'x, ou d'y, ou de
telle autre quantité qu'on aura ſuppoſee , ſera double en
cete equation, c'eſt a dire qu'il y aura deux racines ineſgales entre elles; & dont l'vne ſera C M, l'autre E Q , ſi
c'eſt x qu'on cherche; oubien l'vne ſera M A , & l'autre
Q A, ſi c'eſt y. & ainſi des autres. Il eſt vray que ſi le
point E ne ſe trouue pas du meſme coſté de la courbe
que le point C; il n'y aura que l'vne de ces deux racines
qui ſoit vraye, & l'autre ſera renuerſée, ou moindre que
rien: mais plus ces deux poins, C, & E, ſont proches l'vn
de l'autre, moins il y a de difference entre ces deux racines;

at C but also in another point. Now if this circle cuts CE, the equation involving x and y as unknown quantities (supposing PA and PC known) must have two unequal roots. Suppose, for example, that the circle cuts the curve in the points C and E. Draw EQ parallel to CM. Then x and y may be used to represent EQ and QA respectively in just the same way as they were used to represent CM and MA; since PE is equal to PC (being radii of the same circle), if we seek EQ and QA (supposing PE and PA given) we shall get the same equation that we should obtain by seeking CM and MA (supposing PC and PA given). It follows that the value of x, or y, or any other such quantity, will be two-fold in this equation, that is, the equation will have two unequal roots. If the value of x be required, one of these roots will be CM and the other EQ; while if y be required, one root will be MA and the other QA. It is true that if E is not on the same side of the curve as C, only one of these will be a true root, the other being drawn in the opposite direction, or less than nothing.[150] The nearer together the points C and E are taken however, the less differ-

[150] "Et l'autre sera renversée ou moindre que rien."

ence there is between the roots; and when the points coincide, the roots are exactly equal, that is to say, the circle through C will touch the curve CE at the point C without cutting it.

Furthermore, it is to be observed that when an equation has two equal roots, its left-hand member must be similar in form to the expression obtained by multiplying by itself the difference between the unknown quantity and a known quantity equal to it;[151] and then, if the resulting expression is not of as high a degree as the original equation, multiplying it by another expression which will make it of the same degree. This last step makes the two expressions correspond term by term.

For example, I say that the first equation found in the present discussion,[152] namely

$$y^2 + \frac{qry - 2qvy + qv^2 - qs^2}{q - r},$$

must be of the same form as the expression obtained by making $e=y$ and multiplying $y-e$ by itself, that is, as $y^2-2ey+e^2$. We may then compare the two expressions term by term, thus: Since the first term, y^2, is the same in each, the second term,[153] $\dfrac{qry - 2qvy}{q-r}$, of the first is equal to $-2ey$, the second term of the second; whence, solving for v, or PA, we have $v = e - \dfrac{r}{q}e + \dfrac{1}{2}r$; or, since we have assumed e equal to y, $v = y - \dfrac{r}{q}y + \dfrac{1}{2}r$. In the same way, we can find s from the third term,

[151] That is, the left-hand member will be the square of the binomial $x - a$ when $x = a$.

[152] See page 96. The original has "first equation," not "first member of the equation."

[153] That is, the second term in y.

nes; & enfin elles font entierement efgales, s'ils font tous deux ioins en vn; c'eſt a dire ſi le cercle, qui paſſe par C, y touche la courbe C E ſans la coupper.

De plus il faut conſiderer, que lorſqu'il y a deux raci-nes eſgales en vne equation, elle a neceſſairement la meſme forme, que ſi on multiplie par ſoy meſme la quan-tité qu'on y ſuppoſe eſtre inconnuë moins la quantité connuë qui luy eſt eſgale, & qu'apres cela ſi cete derniere ſomme n'a pas tant de dimenſions que la precedente, on la multiplie par vne autre ſomme qui en ait autant qu'il luy en manque; affin qu'il puiſſe y auoir ſeparement equation entre chaſcun des termes de l'vne , & chaſcun des termes de l'autre.

Comme par exemple ie dis que la premiere equation trouuée cy deſſus, a ſçauoir

$$yy\frac{\mp qry - 2qvy \mp qvv - qss}{q - r}$$ doit auoir la meſme forme que celle qui ſe produiſt en faiſant e eſgal a y, & multipliant $y - e$ par ſoy meſme, d'où il vient $yy - 2ey + ee$, en ſorte qu'on peut comparer ſeparement chaſcun de leurs ter-mes, & dire que puiſque le premier qui eſt yy eſt tout le meſme en l'vne qu'en l'autre, le ſecond qui eſt en l'vne $\frac{qry - 2qvy}{q - r}$ eſt eſgal au ſecõd de l'autre qui eſt $-2ey$, d'où cherchant la quantité v qui eſt la ligne P A , on à

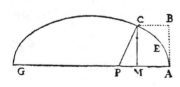

$v \infty e - \frac{r}{q}e + \frac{1}{2}r$, ou biẽ a cauſe que nous auons ſuppoſé e eſgal a y, on a $v \infty y - \frac{r}{q}y + \frac{1}{2}r$. Et

X x 2 ainſi

105

ainſi on pourroit trouuer s par le troiſieſme terme $ee \infty \dfrac{qvv--qss}{q--r}$ mais pourceque la quantité v determine aſſés le point P, qui eſt le ſeul que nous cherchions, on n'a pas beſoin de paſſer outre.

Tout de meſme la ſeconde equation trouuée cy deſſus, a ſçauoir,

$$y^{6} \left.{\substack{--2cd \\ --2by^{5}+bb \\ +dd}}\right\} y^{4}\left.{\substack{+4bcd \\ -2ddv}}\right\} y^{3}\left.{\substack{--2bbcd \\ +ccdd \\ --ddss \\ +ddvv}}\right\} yy--2bccddy+bbccdd.$$

doit auoir meſme forme, que la ſomme qui ſe produiſt lorſqu'on multiplie $yy--2ey+ee$ par

$$y^{4}+fy^{3}+ggyy+hy+k, \text{ qui eſt}$$

$$y^{6}\left.{\substack{+f \\ --2e}}\right\} y^{5}\left.{\substack{+gg \\ --2ef \\ +ee}}\right\} y^{4}\left.{\substack{+h^{3} \\ --2egg \\ +eef}}\right\} y^{3}\left.{\substack{+k^{4} \\ --2eh_{3} \\ +eegg}}\right\} yy\left.{\substack{--2ek^{4} \\ +eeh_{3}}}\right\} y+eek^{4}:$$

defaçon que de ces deux equations i'en tire ſix autres, qui ſeruent a connoiſtre les ſix quantités $f, g, h, k, v, \&\ s$: D'où il eſt fort ayſé a entendre, que de quelque genre, que puiſſe eſtre la ligne courbe propoſée, il vient touſiours par cete façon de proceder autant d'equations, qu'on eſt obligé de ſuppoſer de quantités, qui ſont inconnues. Mais pour demeſler par ordre ces equations, & trouuer enfin la quantité v, qui eſt la ſeule dont on a beſoin, & à l'occaſion de laquelle on cherche les autres: Il faut premierement par le ſecond terme chercher f, la premiere des quantités inconnuës de la derniere ſomme, & on trouue $f \infty 2e--2b$.

Puis par le dernier il faut chercher k la derniere des quantités inconnuës de la meſme ſomme, & on trouue $k^{4} \infty \dfrac{bbccdd}{ee}$

Puis

$e^2 = \dfrac{qv^2 - qs^2}{q - r}$; but since v completely determines P, which is all that is required, it is not necessary to go further.[154]

In the same way, the second equation found above,[155] namely,

$$y^6 - 2by^5 + (b^2 - 2cd + d^2)y^4 + (4bcd - 2d^2v)y^3$$
$$+ (c^2d^2 - 2b^2cd + d^2v^2 - d^2s^2)y^2 - 2bc^2d^2y + b^2c^2d^2,$$

must have the same form as the expression obtained by multiplying

$$y^2 - 2ey + c^2 \text{ by } y^4 + fy^3 + g^2y^2 + h^3y + k^4,$$

that is, as

$$y^6 + (f - 2e)y^5 + (g^2 - 2ef + c^2)y^4 + (h^3 - 2eg^2 + e^2f)y^3$$
$$+ (k^4 - 2eh^3 + c^2g^2)y^2 + (e^2h^3 - 2ek^4)y + e^2k^4.$$

From these two equations, six others may be obtained, which serve to determine the six quantities $f, g, h, k, v,$ and s. It is easily seen that to whatever class the given curve may belong, this method will always furnish just as many equations as we necessarily have unknown quantities. In order to solve these equations, and ultimately to find v, which is the only value really wanted (the others being used only as means of finding v), we first determine f, the first unknown in the above expression, from the second term. Thus, $f = 2e - 2b$. Then in the last terms we can find k, the last unknown in the same expression, from

[154] That is, to construct PC we may lay off $AP = v$ and join P and C. If instead we use the value of e, taking C as center and a radius $CP = e$, we construct an arc cutting AG in P, and join P and C. Rabuel, p. 309. To apply Descartes's method to the circle, for example, it is only necessary to observe that all parameters and diameters are equal, that is, $q = r$; and therefore the equation $v = y - \dfrac{r}{q}y + \dfrac{1}{2}r$ becomes $v = \dfrac{1}{2}q = \dfrac{1}{2}$ diameter. That is, the normal passes through the center and is a radius of the circle. Rabuel, p. 313.

[155] See page 99. As before, Descartes uses "second equation" for "first member of the second equation."

which $k^4 = \dfrac{b^2c^2d^2}{e^2}$. From the third term we get the second quantity

$$g^2 = 3e^2 - 4be - 2cd + b^2 + d^2.$$

From the next to the last term we get h, the next to the last quantity, which is[156]

$$h^3 = \frac{2b^2c^2d^2}{e^3} - \frac{2bc^2d^2}{e^2}.$$

In the same way we should proceed in this order, until the last quantity is found.

Then from the corresponding term (here the fourth) we may find v, and we have

$$v = \frac{2e^3}{d^2} - \frac{3be^2}{d^2} + \frac{b^2e}{d^2} - \frac{2ce}{d} + e + \frac{2bc}{d} + \frac{bc^2}{e^2} - \frac{b^2c^2}{e^3};$$

or putting y for its equal e, we get

$$v = \frac{2y^3}{d^2} - \frac{3by^2}{d^2} + \frac{b^2y}{d^2} - \frac{2cy}{d} + y + \frac{2bc}{d} + \frac{bc^2}{y^2} - \frac{b^2c^2}{y^3},$$

for the length of AP.

[156] Found from.

Puis par le troifiefme terme il faut chercher _g_ la feconde quantité, & on a $gg \infty\ 3\ ee - 4\ be - 2\ cd + bb + dd$.

Puis par le penultiefme il faut chercher _h_ la pénultiefme quantité, qui eft $h \ ; \infty\ \dfrac{2\,bbccdd}{e^3} - \dfrac{2\,bccdd}{ee}$. Et ainfi il faudroit continuer fuiuant ce mefme ordre iufques a la derniere, s'il y en auoit d'auantage en cete fomme ; car c'eft chofe qu'on peut toufiours faire en mefme façon.

Puis par le terme qui fuit en ce mefme ordre , qui eft icy le quatriefme, il faut chercher la quantité _v_ , & on a

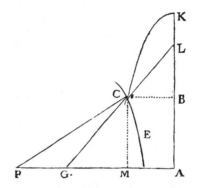

$$v \infty \frac{2\,e^3}{dd} - \frac{3\,bee}{dd} + \frac{bbe}{dd} - \frac{2\,ce}{d} + e + \frac{2\,bc}{d} + \frac{bcc}{ee} - \frac{bbcc}{e^3}.$$

ou mettant _y_ au lieu d'_e_ qui luy eft efgal on a

$$v \infty \frac{2\,y^3}{dd} - \frac{3\,byy}{dd} + \frac{bby}{dd} - \frac{2\,cy}{d} + y + \frac{2\,bc}{d} + \frac{bcc}{yy} - \frac{bbcc}{y^3}.$$

pour la ligne A P.

Et ainfi la troifiefme equation, qui eft

Xx 3   ʑʑ-

$$\frac{\mp 2\,bcddz - 2\,bcdez - 2\,cddvz - 2\,bdevz - bddss \mp bddvv-}{bdd \mp cee \mp eev-}$$

$$\frac{-cddss \mp cddvv,}{-ddv}$$ a la mefme forme que

$z\,z - 2\,f\,z + ff$, en fuppofant f efgal a z, fi bienque il y a derechef equation entre $-2f$, ou $-2z$, &

$$\frac{\mp 2\,bcdd - 2\,bcde - 2\,cddv - 2\,bdev.}{bdd \mp cee \mp eev - ddv}$$ d'où ou connoift que

la quantité v eft $\dfrac{bcdd - bcde \mp bddz \mp ceez}{cdd \mp bde - eez \mp ddz}$

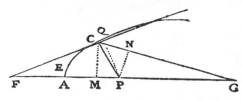

C'eft pourquoy compofant la ligne A P , de cete fomme ef- gale à v dont toutes les quan- tités font connuës, & tirant du point P ainfi trouué, vne ligne droite vers C, elle y couppe la courbe C E a an- gles droits. qui eft ce qu'il falloit faire. Et ie ne voy rien qui empefche, qu'on n'eftende ce problefme en mefme façon a toutes les lignes courbes, qui tombent fous quel- que calcul Geometrique.

Mefme il eft a remarquer touchant la derniere fom- me, qu'on prent a difcretion, pour remplir le nombre des dimenfions de l'autre fomme , lorfqu'il y en man- que , comme nous auons pris tantoft

$y^4 + fy^3 + gg\,yy + h^3 y + k^4$; que les fignes $+$ & $-$ y peuuent eftre fuppofés tels, qu'on veut, fans que la lí- gne v, ou A P, fe trouue diuerfe pour cela, comme vous pourrés ayfement voir par experience. car s'il falloit que ie m'areftaffe a demonftrer tous les theorefmes dont ie

fais

110

Again, the third[107] equation, namely,

$$z^2 + \frac{2bcd^2z - 2bcdez - 2cd^2vz - 2bdevz - bd^2s^2 + bd^2v^2 - cd^2s^2 + cd^2v^2}{bd^2 + ce^2 + e^2v - d^2v},$$

is of the same form as $z^2 - 2fz + f^2$ where $f = z$, so that $-2f$ or $-2z$ must be equal to

$$\frac{2bcd^2 - 2bcde - 2cd^2v - 2bdev}{bd^2 + ce^2 + e^2v - d^2v},$$

whence

$$v = \frac{bcd^2 - bcde + bd^2z + ce^2z}{cd^2 + bde - e^2z + d^2z}.$$

Therefore, if we take AP equal to the above value of v, all the terms of which are known, and join the point P thus determined to C, this line will cut the curve CE at right angles, which was required. I see no reason why this solution should not apply to every curve to which the methods of geometry are applicable.[158]

It should be observed regarding the expression taken arbitrarily to raise the original product to the required degree, as we just now took

$$y^4 + fy^3 + g^2y^2 + h^3y + k^4,$$

that the signs $+$ and $-$ may be chosen at will, without producing different values of v or AP.[159] This is easily found to be the case, but if I should stop to demonstrate every theorem I use, it would require a

[107] First member of the third equation.

[158] Let us apply this method to the problem of constructing a normal to a parabola at a given point. As before, $s^2 = x^2 + v^2 - 2vy + y^2$. If we take as the equation of the parabola $x^2 = ry$, and substitute, we have

$$s^2 = ry + v^2 - 2vy + y^2 \quad \text{or} \quad y^2 + (r - 2v)y + v^2 - s^2 = 0.$$

Comparing this with $y^2 - 2cy + e^2 = 0$, we have $r - 2v = -2e$; $v^2 - s^2 = e^2$; $v = \frac{r}{2} + e$. Since $e = y$, $v = \frac{r}{2} + y$. Let AM $= y$. and $v =$ AP; then AM $-$ AP $=$ MP $=$ one-half the parameter. Rabuel, p. 314.

[159] It will be observed that Descartes did not consider a coefficient, as a, in the general sense of a positive or a negative quantity, but that he always wrote the sign intended. In this sentence, however, he suggests some generalization.

much larger volume than I wish to write. I desire rather to tell you in passing that this method, of which you have here an example, of supposing two equations to be of the same form in order to compare them term by term and so to obtain several equations from one, will apply to an infinity of other problems and is not the least important feature of my general method.[160]

I shall not give the constructions for the required tangents and normals in connection with the method just explained, since it is always easy to find them, although it often requires some ingenuity to get short and simple methods of construction.

[160] The method may be used to draw a normal to a curve from a given point, to draw a tangent to a curve from a point without, and to discover points of inflexion, maxima, and minima. Compare Descartes's Letters, Cousin, Vol. VI, p. 421. As an illustration, let it be required to find a point of inflexion on the first cubical parabola. Its equation is $y^3 = a^2x$. Assume that D is a point of inflexion, and let $CD = y$, $AC = x$, $PA = s$, and $AE = r$. Since triangle PAE is similar to triangle PCD we have $\dfrac{y}{x+s} = \dfrac{r}{s}$, whence $x = \dfrac{sy - rs}{r}$. Substituting in the equation of the curve, we have $y^3 - \dfrac{a^2sy}{r} + a^2s = 0$. But if D is a point of inflexion this equation must have three equal roots, since at a point of inflexion there are three coincident points of section. Compare the equation with

$$y^3 - 3ey^2 + 3e^2y - e^3 = 0.$$

Then $3e^2 = 0$ and $e = 0$. But $e = y$, and therefore $y = 0$. Therefore the point of inflexion is $(0, 0)$. Rabuel, p. 321.

It will be of interest to compare the method of drawing tangents given by Fermat in *Methodus ad disquirendam maximam et minimam*, Toulouse, 1679, which is as follows: It is required to draw a tangent to the parabola BD from a point O without. From the nature of the parabola $\dfrac{CD}{DI} > \dfrac{\overline{BC}^2}{\overline{OI}^2}$, since O is without the curve. But by similar triangles $\dfrac{\overline{BC}^2}{\overline{OI}^2} = \dfrac{\overline{CE}^2}{\overline{IE}^2}$. Therefore $\dfrac{CD}{DI} > \dfrac{\overline{CE}^2}{\overline{IE}^2}$. Let $CE = a$, $CI = e$, and $CD = d$; then $DI = d - e$, and $\dfrac{d}{d-e} > \dfrac{a^2}{(a-e)^2}$; whence

$$de^2 - 2ade > - a^2e.$$

Dividing by e, we have $de - 2ad > - a^2$. Now if the line BO becomes tangent to the curve, the point B and O coincide, $de - 2ad = - a^2$, and e vanishes; then $2ad = a^2$ and $a = 2d$ in length. That is $CE = 2CD$.

fais quelque mention, ie ferois contraint d'eſcrire vn volume beaucoup plus gros que ie ne deſire. Mais ie veux
bien en paſſant vous auertir que l'inuention de ſuppoſer
deux equations de meſme forme, pour comparer ſeparement tous les termes de l'vne a ceux de l'autre, & ainſi
en faire naiſtre pluſieurs d'vne ſeule, dont vous auez vû
icy vn exemple, peut ſeruir a vne infinité d'autres Probleſmes, & n'eſt pas l'vne des moindres de la methode
dont ie me ſers.

Ie n'adiouſte point les conſtruĉtions, par leſquelles on
peut deſcrire les contingentes ou les perpendiculaires
cherchées, en ſuite du calcul que ie viens d'expliquer, a
cauſe qu'il eſt touſiours ayſé de les trouuer: Bienque fouuent on ait beſoin d'vn peu d'adreſſe, pour les rendre
courtes & ſimples.

Comme par exemple, ſi D C eſt la premiere conchoide des anciens,
dont A ſoit le po-
le, & B H la regle:
en ſorte que tou-
tes les lignes droi-
tes qui regardent
vers A, & ſont
compriſes entre la
courbe C D, & la
droite B H, com-

Exemple
de la con-
ſtruĉtion
de ce pro-
bleſme, en
la con-
choide.

me D B & C E, ſoient eſgales: Et qu'on veuille trouuer
la ligne C G qui la couppe au point C a angles droits.
On pourroit en cherchant, dans la ligne B H, le point
par où cete ligne C G doit paſſer, ſelon la methode icy
expli-

expliquée, s'engager dans vn calcul autant ou plus long qu'aucun des precedens: Et toutefois la conftruction, qui deuroit aprés en eftre deduite, eft fort fimple. Car il ne faut que prendre C F en la ligne droite C A, & la faire efgale à C H qui eft perpendiculaire fur H B : puis du point F tirer F G, parallele à B A, & efgale à E A : au moyen de quoy on a le point G, par lequel doit paffer C G la ligne cherchée.

Au refte affin que vous fçachiées que la confideration des lignes courbes icy propofée n'eft pas fans vfage, & qu'elles ont diuerfes proprietés, qui ne cedent en rien a celles des fections coniques, ie veux encore adioufter icy l'explication de certaines Ouales, que vous verrés eftre tres vtiles pour la Theorie de la Catoptrique, & de la Dioptrique. Voycy la facon dont ie les defcris.

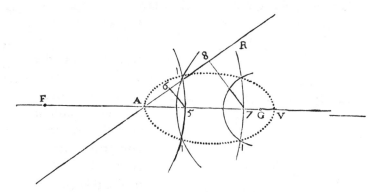

Premierement ayant tiré les lignes droites F A, & A R, qui s'entrecouppent au point A, fans qu'il importe a quels angles, ie prens en l'vne le point F a difcretion, c'eft a dire plus ou moins efloigné du point A felon que

ie

114

Given, for example, CD, the first conchoid of the ancients (see page 113). Let A be its pole and BH the ruler, so that the segments of all straight lines, as CE and DB, converging toward A and included between the curve CD and the straight line BH are equal. Let it be required to find a line CG normal to the curve at the point C. In trying to find the point on BH through which CG must pass (according to the method just explained), we would involve ourselves in a calculation as long as, or longer than any of those just given, and yet the resulting construction would be very simple. For we need only take CF on CA equal to CH, the perpendicular to BH; then through F draw FG parallel to BA and equal to EA, thus determining the point G, through which the required line CG must pass.

To show that a consideration of these curves is not without its use, and that they have diverse properties of no less importance than those of the conic sections I shall add a discussion of certain ovals which you will find very useful in the theory of catoptrics and dioptrics. They may be described in the following way: Drawing the two straight lines FA and AR (p. 114) intersecting at A under any angle, I choose arbitrarily a point F on one of them (more or less distant from A according as the oval is to be large or small). With F as center I describe a circle cutting FA at a point a little beyond A, as at the point 5. I then draw the straight line 56[161] cutting AR at 6, so that A6 is less than A5, and so that A6 is to A5 in any given ratio, as, for example, that which measures the refraction,[162] if the oval is to be used for dioptrics. This being done, I take an arbitrary point G in the line FA on the same side as the point 5, so that AF is to GA in any given ratio. Next, along the line A6 I lay off RA equal to GA, and with G as center and a radius equal to R6 I describe a circle. This circle will cut the first one in two points 1, 1,[163] through which the first of the required ovals must pass.

Next, with F as center I describe a circle which cuts FA as little nearer to or farther from A than the point 5, as, for example, at the point 7. I then draw 78 parallel to 56 and with G as center and a radius equal to R8 I describe another circle. This circle will cut the one through 7 in the points 1, 1[164] which are points of the same oval. We can thus find as many points as may be desired, by drawing lines parallel to 78 and describing circles with F and G as centers.

[161] The confusion resulting from the use of Arabic figures to designate points is here apparent.

[162] That is, the ratio corresponding to the index of refraction.

[163] "Au point 1."

[164] "Au point 1."

ie veux faire ces Ouales plus ou moins grandes, & de ce
point F comme centre ie defcris vn cercle , qui paſſe
quelque peu au delà du point A, comme par le point 5,
puis de ce point 5 ie tire la ligne droite 5 6, qui couppe
l'autre au point 6, en forte qu' A 6 foit moindre qu' A 5,
ſelon telle proportion donnée qu'on veut, a ſçauoir ſelon celle qui meſure les Refractions ſi on s'en veut ſeruir pour la Dioptrique. Aprés cela ie prens auſſy le point
G, en la ligne F A, du coſté où eſt le point 5, a diſcretion,
c'eſt a dire en faiſant que les lignes A F & G A ont entre
elles telle proportion donnée qu'on veut.   Puis ie fais
R A eſgale à G A en la ligne A 6.   & du centre G deſcri-
uant vn cercle, dont le rayon ſoit eſgal à R 6, il couppe
l'autre cercle de part & d'autre au point 1, qui eſt l'vn de
ceux par où doit paſſer la premiere des Ouales cherchées.   Puis derechef du centre F ie defcris vn cercle,
qui paſſe vn peu au deça, ou au delà du point 5, comme
par le point 7, & ayant tiré la ligne droite 7 8 parallele a
5 6, du centre G ie defcris vn autre cercle, dont le rayon
eſt eſgal a la ligne R 8.   & ce cercle couppe celuy qui
paſſe par le point 7 au point 1, qui eſt encore l'vn de ceux
de la meſme Ouale.   Et ainſi on en peut trouuer autant d'autres qu'on voudra , en tirant derechef d'autres lignes paralleles à 7 8, & d'autres cercles des centres
F, & G.

Pour la ſeconde Ouale il n'y a point de difference, ſinon qu'au lieu d' A R il faut de l'autre coſté du point A
prendre A S eſgal à A G, & que le rayon du cercle deſcrit du centre G, pour coupper celuy qui eſt deſcrit du
centre F & qui paſſe par le point 5 , ſoit eſgal a la
Y y ligne

117

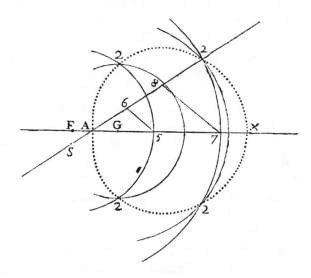

ligne S *6*; ou qu'il foit efgal à S 8 , fi c'eft pour coupper
eeluy qui paſſe par le point 7. & ainſi des autres.　au
moyen dequoy ces cercles s'entrecouppent aux poins
marqués 2, 2 , qui font ceux de cete feconde Ouale
A 2 X.

Pour la troiſiefme, & la quatriefme, au lieu de la ligne
A G il faut prendre A H de l'autre coſté du point A, à
fçauoir du mefme qu'eſt le point F. Et il y a icy de plus
a obferuer que cete ligne A H doit eſtre plus grande que
A F: laquelle peut mefme eſtre nulle , en forte que le
point F fe rencontre où eſt le point A , en la defcription
de toutes ces ouales. Aprés cela les lignes A R , & A S
eſtant efgales à A H , pour defcrire la troiſiefme ouale
A 3 Y, ie fais vn cercle du centre H, dont le rayon eſt
efgal

In the construction of the second oval the only difference is that instead of AR we must take AS on the other side of A, equal to AG, and that the radius of the circle about G cutting the circle about F and passing through 5 must be equal to the line S6; or if it is to cut the circle through 7 it must be equal to S8, and so on. In this way the circles intersect in the points 2, 2, which are points of this second oval A2X.

To construct the third and fourth ovals (see page 121), instead of AG I take AH on the other side of A, that is, on the same side as F. It should be observed that this line AH must be greater than AF, which in any of these ovals may even be zero, in which case F and A coincide. Then, taking AR and AS each equal to AH, to describe the third oval,

A3Y, I draw a circle about H as center with a radius equal to S6 and cutting in the point 3 the circle about F passing through 5, and another with a radius equal to S8 cutting the circle through 7 in the point also marked 3, and so on.

Finally, for the fourth oval, I draw circles about H as center with radii equal to R6, R8, and so on, and cutting the other circles in the points marked 4.[165]

[165] In all four ovals AF and AR or AF and AS intersect at A under any angle. F may coincide with A, and otherwise its distance from A determines the size of the oval. The ratio A5 : A6 is determined by the index of refraction of the material used. In the first two ovals, if A does not coincide with F it lies between F and G, and the ratio AF : AG is arbitrary. In the last two, if F does not coincide with A it lies between A and H, and the ratio AF : AH is arbitrary. In the first oval AR = AG and the points R, 6, 8 are on the same side of A. In the second oval AS = AG and S is on the opposite side of A from 6, 8. In the third oval AS = AH and S is on the opposite side of A from 6, 8. In the fourth oval AR = AH and R, 6, 8 are on the same side of A. Rabuel, p. 342.

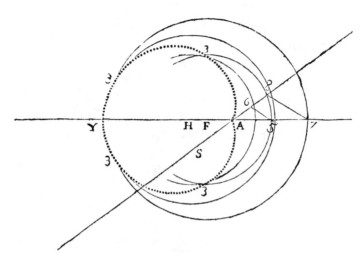

efgal à S 6, qui couppe au point 3 celuy du centre F, qui paffe par le point 5; & vn autre dont le rayon eft efgal a S 8, qui couppe celuy qui paffe par le point 7, au point auffy marqué 3; & ainfi des autres. Enfin pour la derniere

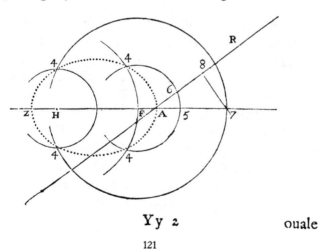

Yy 2   ouale

ouale ie fais des cercles du centre H , dont les rayons
font efgaux aux lignes R 6, R 8, & femblables, qui couppent les autres cercles aux poins marqués 4.

On pourroit encore trouuer vne infinité d'autres
moyens pour defcrire ces mefmes ouales. comme par
exemple, on peut tracer la premiere A V, lorfqu'on fuppofe les lignes F A & A G eftre efgales, fi on diuife la
toute F G au point L, en forte que F L foit a L G, com-

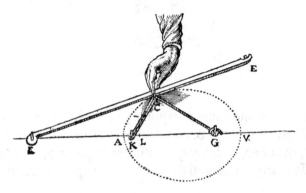

me A 5 à A 6. c'eft à dire qu'elles ayent la proportion,
qui mefure les refractions. Puis ayant diuifé A L en deux
parties efgales au point K, qu'on face tourner vne reigle,
comme F E, autour du point F, en preffant du doigt C,
la chorde E C, qui eftant attachée au bout de cete reigle
vers E, fe replie de C vers K, puis de K derechef vers C,
& de C vers G, ou fon autre bout foit attaché , en forte
que la longeur de cete chorde foit compofée de celle
des lignes G A plus A L plus F E moins A F. & ce fera
le mouuement du point C, qui defcrira cete ouale , a
l'imitation de cequi a efté dit en la Dioptriq; de l'Ellipfe,
　　　　　　　　　　　　　　　　　　　　　&

There are many other ways of describing these same ovals. For example, the first one, AV (provided we assume FA and AG equal) might be traced as follows: Divide the line FG at L so that FL : LG=A5 : A6, that is, in the ratio corresponding to the index of refraction. Then bisecting AL at K, turn a ruler FE about the point F, pressing with the finger at C the cord EC, which, being attached at E to the end of the ruler, passes from C to K and then back to C and from C to G, where its other end is fastened. Thus the entire length of the cord is composed of GA+AL+FE—AF, and the point C will describe the first oval in a way similar to that in which the ellipse and hyperbola are described in *La Dioptrique*.[166] But I cannot give any further attention to this subject.

Athough these ovals seem to be of almost the same nature, they nevertheless belong to four different classes, each containing an infinity of sub-classes, each of which in turn contains as many different kinds as does the class of ellipses or of hyperbolas; the sub-classes depending upon the value of the ratio of A5 to A6. Then, as the ratio of AF to AG, or of AF to AH changes, the ovals of each sub-class change in kind, and the length of AG or AH determines the size of the oval.[167]

If A5 is equal to A6, the ovals of the first and third classes become straight lines; while among those of the second class we have all possible hyperbolas, and among those of the fourth all possible ellipses.[168]

In the case of each oval it is necessary further to consider two portions having different properties. In the first oval the portion toward A (see page 114) causes rays passing through the air from F to converge towards G upon meeting the convex surface 1A1 of a lens whose index of refraction, according to dioptrics, determines such ratios as that of A5 to A6, by means of which the oval is described.

[166] See the notes on pages 10, 55, 112.

[167] Compare the changes in the ellipse and hyperbola as the ratio of the length of the transverse axis to the distance between the foci changes.

[168] These theorems may be proved as follows: (1) Given the first oval, with A5 = A6; then RA = GA; FP = F5; GP = R6 = AR — R6 = GA — A5 = G5. Therefore FP + GP = F5 + G5. That is, the point P lies on the straight line FG. (2) Given the second oval, with A5 = A6; then F2 = F5 = FA + A5; G2 = S6 = SA + A6 = SA + A5; G2 — F2 = SA — FA = GA — FA = C. Therefore 2 lies on a hyperbola whose foci are F and G, and whose transverse axis is GA — FA. The proof for the third oval is analogous to (1) and that for the fourth to (2).

It may be noted that the first oval is the same curve as that described on page 98. For FP = F5, whence FP — AF = A5, and AR = AG; GP = R6; AG — GP = A6. If then A5 : A6 = d : e we have, as before,

$$FP — AF : AG — GP = d : e.$$

& de l'Hyperbole. mais ie ne veux point m'arefter plus
long tems fur ce fuiet.

Or encore que toutes ces oüales femblent eftre quafi
de mefme nature, elles font neanmoins de 4 diuers genres, chafcun defquels contient fous foy vne infinité d'autres genres, qui derechef contienent chafcun autant de
diuerfes efpeces, que fait le genre des Ellipfes, ou celuy
des Hyperboles. Car felon que la proportion, qui eft entre les lignes A 5, A 6, ou femblables, eft differente ; le
genre fubalterne de ces oüales eft different. Puis felon
que la proportion, qui eft entre les lignes A F, & A G, ou
A H, eft changée, les oüales de chafque genre fubalterne changent d'efpece. Et felon qu' A G, ou A H eft plus
ou moins grande, elles font diuerfes en grandeur. Et fi
les lignes A 5 & A 6 font efgales, au lieu des oüales du
premier genre ou du troifiefme, on ne defcrit que des
lignes droites; mais au lieu de celles du fecond on a toutes les Hyperboles poffibles; & au lieu de celles du dernier toutes les Ellipfes.

Outre cela en chafcune de ces oüales il faut confiderer
deux parties, qui ont diuerfes proprietés ; a fçauoir en la
premiere, la partie qui eft vers A, fait que les rayons, qui
eftant dans l'air vienent du point F, fe retoùruent tous
vers le point G, lorfqu'ils rencontrent la fuperficie conuexe d'vn verre, dont la fuperficie eft 1 A 1, & dans lequel les refractions fe font telles, que fuiuant ce qui a
efté dit en la Dioptrique, elles peuuent toutes eftre me-
furées par la proportion, qui eft entre les lignes A 5 &
A 6, ou femblables, par l'ayde defquelles on a defcrit cete
oüale.

*Les pro-
prietés de
ces oüales
touchant
les refle-
xions, &
les refra-
ctions.*

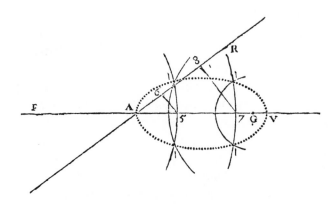

Mais la partie, qui eſt vers V, fait que les rayons qui
vienent du point G ſe reflefchiroient tous vers F, s'ils y
rencontroient la ſuperficie concaue d'vn miroir, dont la
figure fuſt 1 V 1, & qui fuſt de telle matiere qu'il di-
minuaſt la force de ces rayons, ſelon la proportion qui eſt
entre les lignes A 5 & A 6 : Car de ce qui a eſté demon-
ſtré en la Dioptrique, il eſt euident que cela poſé, les angles de la reflexion ſeroient ineſgaus, auſſy bien que ſont
ceux de la refraction, & pourroient eſtre meſurés en
meſme ſorte.

En la ſeconde ouale la partie 2 A 2 ſert encore pour les
reflexions dont on ſuppoſe les angles eſtre ineſgaux. car
eſtant en la ſuperficie d'vn miroir compoſé de meſme
matiere que le precedent, elle feroit tellement reflefchir
tous les rayons, qui viendroient du point G, qu'ils ſem-
bleroient aprés eſtre reflefchis venir du point F. Et il
eſt a remarquer, qu'ayant fait la ligne A G beaucoup
                                                            plus

But the portion toward V causes all rays coming from G to converge toward F when they strike the concave surface of a mirror of the shape of 1V1 and of such material that it diminishes the velocity of these rays in the ratio of A5 to A6, for it is proved in dioptrics that in this case the angles of reflection will be unequal as well as the angles of refraction, and can be measured in the same way.

Now consider the second oval. Here, too, the portion 2A2 (see page 118) serves for reflections of which the angles may be assumed unequal. For if the surface of a mirror of the same material as in the case of the first oval be of this form, it will reflect all rays from G, making them seem to come from F. Observe, too, that if the line AG is considerably greater than AF, such a mirror will be convex in the center (toward A) and concave at each end; for such a curve would be heart-shaped rather than oval. The other part, X2, is useful for refracting lenses; rays which pass through the air toward F are refracted by a lens whose surface has this form.

The third oval is of use only for refraction, and causes rays traveling through the air toward F (page 121) to move through the glass toward H, after they have passed through the surface whose form is A3Y3, which is convex throughout except toward A, where it is slightly concave, so that this curve is also heart-shaped. The difference between the two parts of this oval is that the one part is nearer F and farther from H, while the other is nearer H and farther from F.

Similarly, the last of these ovals is useful only in the case of reflection. Its effect is to make all rays coming from H (see the second figure on page 121) and meeting the concave surface of a mirror of the same material as those previously discussed, and of the form A4Z4, converge towards F after reflection.

The points F, G and H may be called the "burning points" [169] of these ovals, to correspond to those of the ellipse and hyperbola, and they are so named in dioptrics.

I have not mentioned several other kinds of reflection and refraction that are effected[170] by these ovals; for being merely reverse or opposite effects they are easily deduced.

[169] That is, the foci, from the Latin *focus*, "hearth." The word *focus* was first used in the geometric sense by Kepler, *Ad Vitellionem Paralipomena*, Frankfort, 1604. Chap. 4, Sect. 4.

[170] "Reglées."

plus grande que A F, ce miroir feroit conuexe au milieu,
vers A, & concaue aux extrèmitez: car telle eſt la figure
de cete ligne, qui en cela repreſente plutoſt vn coeur
qu'vne ouale.

Mais ſon autre partie X 2 ſert pour les refractions, &
fait que les rayons, qui eſtant dans l'air tendent vers F, ſe
detournent vers G, en trauerſant la ſuperficie d'vn verre, qui en ait la figure.

La troiſieſme ouale ſert toute aux refractions, & fait
que les rayons, qui eſtant dans l'air tendent vers F, ſe
vont rendre vers H dans le verre, aprés qu'ils ont trauer-
ſé ſa ſuperficie, dont la figure eſt A 3 Y 3, qui eſt conuexe par tout, excepté vers A où elle eſt vn peu concaue, en
ſorte qu'elle a la figure d'vn coeur auſſy bien que la precedente. Et la difference qui eſt entre les deux parties
de cete ouale, conſiſte en ce que le point F eſt plus proche de l'vne, que n'eſt le point H; & qu'il eſt plus
eſloigné de l'autre, que ce meſme point H.

En meſme façon la derniere ouale ſert toute aux reflexions, & fait que ſi les rayons, qui vienent du point H,
rencontroient la ſuperficie concaue d'vn miroir de meſme matiere que les precedens, & dont la figure fuſt A 4
Z 4, ils ſe reflefchiroient tous vers F.

De façon qu'on peut nommer les poins F, & G, ou H
les poins bruſlans de ces ouales, a l'exemple de ceux des
Ellipſes, & des Hyperboles, qui ont eſté ainſi nommés
en la Dioptrique.

I'omets quantité d'autres refractions, & reflexions,
qui ſont reiglées par ces meſmes ouales : car n'eſtant
que les conuerſes, ou les contraires de celles cy, elles en
peuuent

Demou-
ſtration
des pro-
prietés de
ces ouales
touchant
les refle-
xions &
refra-
ctions,

peuuent facilement eſtre deduites.　Mais il ne faut pas
que i'omette la demonſtration de ceque iay dit.　& a cet
effect, prenons par exemple le point C a diſcretion en la
premiere partie de la premiere de ces ouales ; puis tirons
la ligne droite
C P, qui coup-
pe la courbe au
point C à an-
gles droits, ce-
qui eſt facile

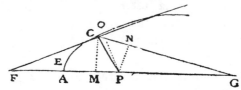

par le probleſme precedent ; Car prenant *b* pour A G, *c*
pour A F, *c* + *z* pour F C; & ſuppoſant que la proportion qui eſt entre *d* & *e*, que ie prendray icy touſiours
pour celle qui meſure les refractions du verre propoſé,
deſigne auſſy celle qui eſt entre les lignes A 5, & A 6, ou
ſemblables, qui ont ſerui pour deſcrire cete ouale, ce qui

donne $b - \frac{e}{d} z$ pour G C: on trouue que la ligne **A** P eſt

$$\frac{bcdd - bcde + bddz + ceez.}{bde + cdd + ddz \div eez}$$ ainſi qu'il a eſté monſtré cy deſſus.

De plus du point P ayant tiré P Q a angles droits ſur la
droite F C, & P N auſſy a angles droits ſur G C, conſide-
rons que ſi P Q eſt à P N, comme *d* eſt à *e*, c'eſt à dire,
comme les lignes qui meſurent les refractions du verre
conuexe A C, le rayon qui vient du point F au point **C**,
doit tellement s'y courber en entrant dans ce verre, qu'il
s'aille rendre aprés vers G: ainſi qu'il eſt tres euident de
cequi a eſté dit en la Dioptrique.　Puis enfin voyons par
le calcul, s'il eſt vray, que P Q ſoit à P N ; comme *d* eſt
à *e*.　Les triangles rectangles P Q F, & C M F ſont ſemblables;

I must not, however, fail to prove the statements already made. For this purpose, take any point C on the first part of the first oval, and draw the straight line CP normal to the curve at C. This can be done by the method given above,[171] as follows:

Let AG=b, AF=c, FC=$c+z$. Suppose the ratio of d to e, which I always take here to measure the refractive power of the lens under consideration, to represent the ratio of A5 to A6 or similar lines used to describe the oval. Then

$$GC = b - \frac{e}{d} z,$$

whence

$$AP = \frac{bcd^2 - bcde + bd^2z + ce^2z}{bde + cd^2 + d^2z - e^2z}.$$

From P draw PQ perpendicular to FC, and PN perpendicular to GC.[172] Now if PQ : PN=d : e, that is, if PQ : PN is equal to the same ratio as that between the lines which measure the refraction of the convex glass AC, then a ray passing from F to C must be refracted toward G upon entering the glass. This follows at once from dioptrics.

[171] See page 115.

[172] Here PQ is the sine of the angle of incidence and PN is the sine of the angle of refraction. The ray FC is reflected along CG.

Now let us determine by calculation if it be true that PQ : PN$=d$: e.

The right triangles PQF and CMF are similar, whence it follows that CF : CM$=$FP : PQ, and $\dfrac{FP.CM}{CF}=PQ$. Again, the right triangles PNG and CMG are similar, and therefore $\dfrac{GP.CM}{CG}=PN$. Now since the multiplication or division of two terms of a ratio by the same number does not alter the ratio, if $\dfrac{FP.CM}{CF}:\dfrac{GP.CM}{CG}=d:e$, then, dividing each term of the first ratio by CM and multiplying each by both CF and CG, we have FP.CG : GP.CF$=d$: e. Now by construction,

$$FP=c+\frac{bcd^2-bcde+bd^2z+cc^2z}{cd^2+bde-e^2z+d^2z},$$

or

$$FP=\frac{bcd^2+c^2d^2+bd^2z+cd^2z}{cd^2+bde-e^2z+d^2z},$$

and

$$CG=b-\frac{e}{d}z.$$

Then

$$FP.CG=\frac{b^2cd^2+bc^2d^2+b^2d^2z+bcd^2z-bcdez-c^2dez-bdez^2-cdez^2}{cd^2+bde-e^2z+d^2z}.$$

Then

$$GP=b-\frac{bcd^2-bcde+bd^2z+ce^2z}{cd^2+bde-e^2z+d^2z};$$

or

$$GP=\frac{b^2de+bcde-be^2z-ce^2z}{cd^2+bde-e^2z+d^2z};$$

and CF$=c+z$. So that

$$GP.CF=\frac{b^2cde+bc^2de+b^2dez+bcdez-bce^2z-c^2e^2z-be^2z^2-ce^2z^2}{cd^2+bde-e^2z+d^2z}.$$

blables; d'où il fuit que C F eft à C M, comme F P eft a
P Q; & par confequent que F P, eftant multipliée par
C M, & diuifée par C F, eft'efgale a P Q. Tout de mefme les triangles rectangles P N G, & C M G font femblables; d'où il fuit que G P, multipliée par C M, & diuifée par C G, eft efgale a P N. Puis a caufe que les mul-
tiplications, ou diuifions, qui fe font de deux quantités
par vne mefme, ne changent point la proportion qui eft
entre elles; fi F P multipliée par C M; & diuifée par C F,
eft à G P multipliée auffy par C M & diuifée par C G;
comme d eft à e, en diuifant l'vne &l'autre de ces deux
fommes par C M , puis les multipliant toutes deux par
C F, & derechef par C G, il refte F P multipliée par C G,
qui doit eftre à G P multipliée par C F, comme d eft à e.

Or par la conftruction F P eft $c \; \dfrac{\pm bcdd -- bcde \pm bddz \pm ceez,}{bde \pm cdd \pm ddz -- eez}$

oubien F P $\infty \dfrac{bcdd \pm ccdd \pm bddz \pm cddz.}{bde \pm cdd \pm ddz -- eez}$ & C G eft

$b -- \dfrac{e}{d} z.$ fibienque multipliant F P par C G il vient

$$\dfrac{bbcdd \pm bccdd \pm bbddz \pm bcddz -- bcdez -- ccdez -- bdezz -- cdezz.}{bde \pm cdd \pm ddz -- eez}$$

Puis G P eft $b \; \dfrac{-- bcdd \pm bcde -- bddz -- ceez.}{bde \pm cdd \pm ddz -- eez}$ oubien

G P $\infty \dfrac{bbde \pm bcde -- beez -- ceez,}{bde \pm cdd \pm ddz -- eez}$ & C F eft $c + z$;

fibienque multipliant G P par C F, il vient

$$\dfrac{bbcde \pm bccde -- bceez -- cceez \pm bbdez \pm bcdez -- beezz -- ceezz.}{bde \pm cdd \pm ddz -- eez}$$

Et pourceque la premiere de ces fommes diuifée par d,
eft la mefme que la feconde diuifée par e, il eft manifefte,
que F P multipliée par C G eft a G P multipliée par C F;

Z z   c'eft

c'eſt a dire que P Q eſt à P N, comme *d* eſt à *e* , qui eſt
tout ce qu'il falloit demonſtrer.

Et ſçachés , que cete meſme demonſtration s'eſtend
a tout cequi a eſté dit des autres refraĉtions ou reflexions, qui ſe font dans les oüales propoſées; ſans qu'il y
faille changer aucune choſe , que les ſignes ✚ & -- du
calcul. c'eſt pourquoy chaſcun les peut ayſement examiner de ſoymeſme, ſans qu'il ſoit beſoin que ie m'y
areſte.

Mais il faut maintenent, que ie ſatisface a ce que iay
omis en la Dioptrique, lorſqu'aprés auoir remarqué qu'il
peut y auoir des verres de pluſieurs diuerſes figures , qui
facent auſſy bien l'vn que l'autre, que les rayons venans
d'vn meſme point de l'obiet, s'aſſemblent tous en vn autre point aprés les auoir trauerſés. & qu'entre ces verres,
ceux qui ſont fort conuexes d'un coſté, & concaues de
l'autre, ont plus de force pour bruſler, que ceux qui ſont
eſgalement conuexes des deux coſtés. au lieu que tout
au contraire ces derniers ſont les meilleurs pour les lunetes. ie me ſuis contente d'expliquer ceux , que i'ay crû
eſtre les meilleurs pour la prattique, en ſuppoſant la diffi-
culté que les artiſans peuuent auoir a les tailler. C'eſt
pourquoy, affin qu'il ne reſte rien a ſouhaiter touchant la
theorie de cete ſcience, ie doy expliquer encore icy la figure des verres, qui ayant l'vne de leurs ſuperficies autant conuexe, ou concaue, qu'on voudra, ne laiſſent pas
de faire que tous les rayons , qui vienent vers eux d'vn
meſme point , ou paralleles , s'aſſemblent aprés en vn
meſme point; & celle des verres qui font le ſemblable,
eſtant eſgalement conuexes des deux coſtés , ou bien la

conue-

The first of these products divided by d is equal to the second divided by e, whence it follows that $PQ : PN = FP.CG : GP.CF = d : e$, which was to be proved. This proof may be made to hold for the reflecting and refracting properties of any one of these ovals, by proper changes of the signs plus and minus; and as each can be investigated by the reader, there is no need for further discussion here.[173]

It now becomes necessary for me to supplement the statements made in my Dioptrique[174] to the effect that lenses of various forms serve equally well to cause rays coming from the same point and passing through them to converge to another point; and that among such lenses those which are convex on one side and concave on the other are more powerful burning-glasses than those which are convex on both sides; while, on the other hand, the latter make the better telescopes.[175] I shall describe and explain only those which I believe to have the greatest practical value, taking into consideration the difficulties of cutting. To complete the theory of the subject, I shall now have to describe

[173] To obtain the equation of the first oval we may proceed as follows: Let $AF = c$; $AG = b$; $FC = c + z$; $GC = b - \frac{e}{d}z$. Let $CM = x$, $AM = y$. $FM = c + y$; $GM = b - y$. Draw PC normal to the curve at any point C. Let $AP = v$. Then $\overline{CF}^2 = \overline{CM}^2 + \overline{FM}^2$. Also, $c^2 + 2cz + z^2 = x^2 + c^2 + 2cy + y^2$, whence

$$z = -c + \sqrt{x^2 + c^2 + 2cy + y^2}.$$

Also, $\overline{CG}^2 = \overline{CM}^2 + \overline{GM}^2$, whence

$$b^2 - 2\frac{be}{d}z + \frac{e^2}{d^2}z^2 = x^2 + b^2 - 2by + y^2.$$

Substituting in this equation the value of z obtained above, squaring, and simplifying, we obtain:

$$\left[(d^2 - e^2)x^2 + (d^2 - e^2)y^2 - 2(e^2c + bd^2)y - 2ec(ec - bd) \right]^2$$
$$= 4e^2(bd + ec)^2(x^2 + c^2 + 2cy + y^2).$$ Rabuel, p. 348.

[174] Descartes: *La Dioptrique*, published with *Discours de la Methode*, Leyden, 1637. See also Cousin, vol. III, p. 401.

[175] "Lunetes." The laws of reflection were familiar to the geometers of the Platonic school, and burning-glasses, in the form of spherical glass shells filled with water, or balls of rock crystal are discussed by Pliny, Hist. Nat. xxxvi, 67 (25) and xxxvii, 10. Ptolemy, in his treatise on Optics, discussed reflection, refraction, and plane and concave mirrors.

again the form of lens which has one side of any desired degree of convexity or concavity, and which makes all the rays that are parallel or that come from a single point converge after passing through it; and also the form of lens having the same effect but being equally convex on both sides, or such that the convexity of one of its surfaces bears a given ratio to that of the other.

In the first place, let G, Y, C, and F be given points, such that rays coming from G or parallel to GA converge at F after passing through a concave lens. Let Y be the center of the inner surface of this lens and C its edge, and let the chord CMC be given, and also the altitude of the arc CYC. First we must determine which of these ovals can be used for a lens that will cause rays passing through it in the direction of H (a point as yet undetermined) to converge toward F after leaving it.

There is no change in the direction of rays by means of reflection or refraction which cannot be effected by at least one of these ovals; and it is easily seen that this particular result can be obtained by using either part of the third oval, marked 3A3 or 3Y3 (see page 121), or the part of the second oval marked 2X2 (see page 118). Since the same method applied to each of these, we may in each case take Y

conuexité de l'vne de leurs fuperficies ayant la proportion donnée à celle de l'autre.

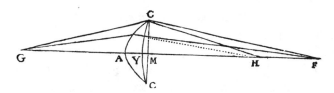

Pofons pour le premier cas, que les poins G, Y, C, & F eftant donnés, les rayons qui vienent du point G, oubien qui font paralleles à G A fe doiuent affembler au point F, aprés auoir trauerfé vn verre fi concaue, qu'Y eftant le milieu de fa fuperficie interieure, l'extremité en foit au point C, en forte que la chorde C M C, & la fleche Y M de l'arc C Y C, font données. La queftion va là, que premierement il faut confiderer, de laquelle des ouales expliquées, la fuperficie du verre Y C, doit auoir la figure, pour faire que tous les rayons, qui eftant dedans tendent vers vn mefme point, comme vers H, qui n'eft pas encore connu, s'aillent rendre vers vn autre, a fçauoir vers F, aprés en eftre fortis. Car il n'y a aucun effect touchant le rapport des rayons changé par reflexion, ou refraction d'vn point a vn autre, qui ne puiffe eftre caufé par quelqu'vne de ces ouales. & on voit ayfement que ceruycy le peut eftre par la partie de la troifiefme Ouale, qui a tantoft efté marquée 3 A 3, ou par celle de la mefme, qui a efté marquée 3 Y 3, ou enfin par la partie de la feconde qui a efté marquée 2 X 2. Et pourceque ces trois tombent icy fous mefme calcul, on doit tant pour l'vne, que pour l'autre prendre Y pour

Commét on peut faire vn verre autant conuexe ou concaue, en l'vne de fes fuperficies, qu'on voudra, qui raffemble a vn point donné, tous les rayons qui vienent d'vn autre point donné.

Zz 2   leur

leur fommet, C pour l'vn des poins de leur circonference, & F pour l'vn de leurs poins bruſlans ; aprés quoy il
ne reſte plus a chercher que le point H, qui doit eſtre
l'autre point bruſlant. Et on le trouue en conſiderant,
que la difference, qui eſt entre les lignes F Y & F C, doit
eſtre a celle, qui eſt entre les lignes H Y & H C, comme
d eſt à e, c'eſt a dire, comme la plus grande des lignes qui
meſurent les refractions du verre propoſé eſt à la moindre; ainſi qu'on peut voir manifeſtement de la deſcription de ces ouales. Et pourceque les lignes F Y & F C
font données, leur difference l'eſt auſſy, & en ſuite celle
qui eſt entre H Y & H C; pourceque la proportion qui
eſt entre ces deux differences eſt donnée. Et de plus a
cauſe que Y M eſt donnée, la difference qui eſt entre
M H, & H C, l'eſt auſſy; & enfin pourceque C M eſt donnée, il ne reſte plus qu'à trouuer M H le coſté du triangle

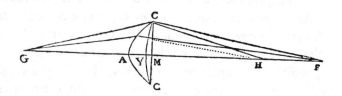

rectangle C M H, dont on a l'autre coſté C M, & on a
auſſy la difference qui eſt entre C H la baze, & M H le
coſté demandé. d'où il eſt ayſé de le trouuer. car ſi on
prent k pour l'excés de C H ſur M H, & n pour la longeur
de la ligne C M, on aura $\frac{nn}{2k} - \frac{1}{2}k$ pour M H. Et aprés
auoir ainſi le point H, s'il ſe trouue plus loin du point Y,
que

(see pages 137 and 138), as the vertex, C as a point on the curve,[176] and F as one of the foci. It then remains to determine H, the other focus. This may be found by considering that the difference between FY and FC is to the difference between HY and HC as d is to e; that is, as the longer of the lines measuring the refractive power of the lens is to the shorter, as is evident from the manner of describing the ovals.

Since the lines FY and FC are given we know their difference; and then, since the ratio of the two differences is known, we know the difference between HY and HC.

Again, since YM is known, we know the difference between MH and HC, and therefore CM. It remains to find MH, the side of the right triangle CMH. The other side of this triangle, CM, is known, and also the difference between the hypotenuse, CH and the required side, MH. We can therefore easily determine MH as follows:

Let $k = CH - MH$ and $n = CM$; then $\dfrac{n^2}{2k} - \dfrac{1}{2}k = MH$, which determines the position of the point H.

[176] "Circonference."

If HY is greater than HF, the curve CY must be the first part of the third class of oval, which has already been designated by 3A3.

But suppose that HY is less than FY. This includes two cases: In the first, HY exceeds HF by such an amount that the ratio of their difference to the whole line FY is greater than the ratio of e, the smaller of the two lines that represent the refractive power, to d, the larger; that is, if $HF=c$, and $HY=c+h$, then dh is greater than $2ce+eh$. In this case CY must be the second part 3Y3 of the same oval of the third class.

In the second case dh is less than or equal to $2ce+eh$, and CY is the second part 2X2 of the oval of the second class.

Finally, if the points H and F coincide, $FY = FC$ and the curve YC is a circle.

It is also necessary to determine CAC, the other surface of the lens. If we suppose the rays falling on it to be parallel, this will be an ellipse having H as one of its foci, and the form is easily determined. If, however, we suppose the rays to come from the point G, the lens must have the form of the first part of an oval of the first class, the two foci of which are G and H and which passes through the point C. The point A is seen to be its vertex from the fact that the excess of GC over GA is to the excess of HA over HC as d is to e. For if k represents the difference between CH and HM, and x represents AM, then $x-k$ will represent the difference between AH and CH; and if g represents the difference between GC and GM, which are given, $g+x$

que n'en eſt le point F, la ligne C Y doit eſtre la premiere partie de l'ouale du troiſieſme genre, qui a tantoſt eſté
nommée 3 A 3: Mais ſi H Y eſt moindre que F Y, oubien
elle ſurpaſſe H F de tant, que leur difference eſt plus
grande a raiſon de la toute F Y, que n'eſt *e* la moindre
des lignes qui meſurent les refractions comparée auec *d*
la plus grande, c'eſt a dire que faiſant H F ∞ *c*, &
H Y ∞ *c* + *h*, *dh* eſt plus grande que 2 *ce* + *eh*, & lors
C Y doit eſtre la ſeconde partie de la meſme ouale du
troiſieſme genre, qui a tantoſt eſté nomée 3 Y 3; Oubien
dh eſt eſgale, ou moindre que 2 *ce* + *eh*; & lors C Y
doit eſtre la ſeconde partie de l'ouale du ſecond genre
qui a cy deſſus eſté nommée 2 X 2. Et enfin ſi le point H
eſt le meſme que le point F, ce qui n'arriue que lorſque
F Y & F C ſont eſgales cete ligne Y C eſt vn cercle.

Aprés cela il faut chercher C A C l'autre ſuperficie de
ce verre, qui doit eſtre vne Ellipſe, dont H ſoit le point
bruſlant; ſi on ſuppoſe que les rayons qui tombent deſſus
ſoiēt paralleles; & lors il eſt ayſé de la trouuer. Mais ſi on
ſuppoſe qu'ils vienēt du point G, ce doit eſtre la premiere
partie d'vne ouale du premier genre, dont les deux poins
bruſlans ſoiēt G & H, & qui paſſe par le point C: d'où on
trouue le point A pour le ſommet de cete ouale, en conſi-
derāt, que G C doit eſtre plus grāde que G A, d'vne quantité, qui ſoit a celle dont H A ſurpaſſe H C, comme *d* à *e*.
car ayant pris *k* pour la difference, qui eſt entre C H, & H
M, ſi on ſuppoſe *x* pour A M, on aura *x* -- *k*, pour la difference qui eſt entre A H, & C H; puis ſi on prent *g* pour
celle, qui eſt entre G C, & G M, qui ſont données, on
aura *g* + *x* pour celle, qui eſt entre G C, & G A ; &

Z z 3   pour-

Commēt
on peut
faire vn
verre, qui
ait le mef-
me effect
que le
precedēt,
& que la
conuexi-
té del'vne
de fes fu-
perficies
ait la pro-
portion
donnée
auec celle
del'autre.

pourceque cete derniere $g + x$ eft à l'autre $x - k$, com-
me d eft à e. on à $ge + ex \infty dx - dk$, oubien $\frac{ge + dk}{d - e}$
pour la ligne x, ou A M , par laquelle on determine le
point A qui eftoit cherché.

Pofons maintenent pour l'autre cas, qu'on ne donne
que les poins G C, & F, auec la proportion qui eft entre
les lignes A M, & Y M, & qu'il faille trouuer la figure du
verre A C Y, qui face que tous les rayons, qui vienent
du point G s'affemblent au point F.

On peut derechef icy fe feruir de deux ouales dont
l'vne, A C, ait G & H pour fes poins bruflans; & l'autre,

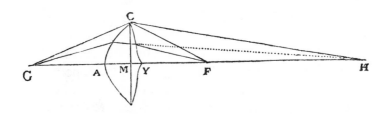

C Y, ait F & H pour les fiens. Et pour les trouuer, premierement fuppofant le point H qui eft commun a toutes
deux eftre connu, ie cherche A M par les trois poins
G, C, H, en la façon tout maintenent expliquée; a fçauoir
preuant k pour la difference, qui eft entre C H, & H M;
& g pour celle qui eft entre G C, & G M : & A C eftant
la premiere partie de l'Ouale du premier genre, iay
$\frac{ge + dk}{d - e}$ pour A M: puis ie cherche auffy M Y par les trois
poins F, C, H, en forte que C Y foit la premiere partie
d'vne ouale du troifiefme genre; & prenant y pour M Y,
&

will represent the difference between GC and GA; and since $g+x : x-k=d : e$, we have $ge+ex=dx-dk$, or $AM=x=\dfrac{ge+dk}{d-e}$, which enables us to determine the required point A.

Again, suppose that only the points G, C, and F are given, together with the ratio of AM to YM; and let it be required to determine the form of the lens ACY which causes all the rays coming from the point G to converge to F.

In this case, we can use two ovals, AC and CY, with foci G and H, and F and H respectively. To determine these let us suppose first that H, the focus common to both, is known. Then AM is determined by the three points G, C, and H in the way just now explained; that is if k represents the difference between CH and HM, and g the difference between GC and GM, and if AC be the first part of the oval of the

first class, we have $AM=\dfrac{ge+dk}{d-e}$.

We may then find MY by means of the three points F, C, and H. If CY is the first part of an oval of the third class and we take y for MY and f for the difference between CF and FM, we have the difference between CF and FY equal to $f+y$; then let the difference between CH and HM equal k, and the difference between CH and HY equal $k+y$. Now $k+y : f+y = e : d$, since the oval is of the third class, whence $MY = \dfrac{fe-dk}{d-e}$. Therefore, $AM+MY = AY = \dfrac{ge+fe}{d-e}$, whence it follows that on whichever side the point H may lie, the ratio of the line AY to the excess of GC+CF over GF is always equal to the ratio of e, the smaller of the two lines representing the refractive power of the glass, to $d-e$, the difference of these two lines, which gives a very interesting theorem.[177]

The line AY being found, it must be divided in the proper ratio into AM and MY, and since M is known the points A and Y, and finally the point H, may be found by the preceding problem. We must first find whether the line AM thus found is greater than, equal to, or less

than $\dfrac{ge}{d-e}$. If it is greater, AC must be the first part of one of the third class, as they have been considered here. If it is smaller, CY must be the first part of an oval of the first class and AC the first part

[177] "Qui est un assez beau théorème."

&_f_ pour la difference, qui eſt entre C F, & F M , i'ay
_f+y_, pour celle qui eſt entre C F, & F Y: puis ayant deſia _k_ pour celle qui eſt entre C H, & H M, iay _k_ +_y_ pour
celle qui eſt entre C H, & H Y, que ie ſcay deuoir eſtre
à _f+y_ comme _e_ eſt à _d_, a cauſe de l'Ouale du troiſieſme
genre, d'où ie trouue que _y_ ou M Y eſt $\frac{fe\cdot\cdot dk,}{d\cdot\cdot e}$ puis ioignant enſemble les deux quantités trouuées pour A M, &
M Y, ie trouue $\frac{ge+fe}{d\cdot\cdot e}$ pour la toute A Y; D'où il ſuit que
de quelque coſté que ſoit ſuppoſé le point H, cete ligne
A Y eſt touſiours compoſée d'vne quantité, qui eſt a celle dont les deux enſemble G C, & C F ſurpaſſent la toute G F, Comme _e_, la moindre des deux lignes qui ſeruent
a meſurer les refractions du verre propoſé, eſt à _d--e_, la
difference qui eſt entre ces deux lignes. ce qui eſt vn aſſés beau theoreſme. Or ayant ainſi la toute A Y, il la
faut couper ſelon la proportion que doiuent auoir ſes
parties A M & M Y; au moyen de quoy pource qu'on a
deſia le point M, on trouue auſſy les poins A & Y; & en
ſuite le point H, par le probleſme precedent. Mais auparauant il faut regarder, ſi la ligne A M ainſi trouuée eſt
plus grande que $\frac{ge,}{d\cdot\cdot e}$ ou plus petite, ou eſgale. Car ſi elle
eſt plus grande, on apprent de là que la courbe A C doit
eſtre la premiere partie d'vne ouale du premier genre; &
C Y la premiere d'vne du troiſieſme, ainſi qu'elles ont
eſté icy ſuppoſées: au lieu que ſi elle eſt plus petite, cela
monſtre que c'eſt C Y, qui doit eſtre la premiere partie
d'vne ouale du premier genre; & que A C doit eſtre la
premiere d'vne du troiſieſme : Enfin ſi A M eſt eſgale à
$$\frac{ge,}{d-e}$$

$\frac{g\,e,}{d\,--e}$ les deux courbes A C & C Y doiuent eſtre deux hyperboles.

On pourroit eſtendre ces deux probleſmes a vne infinité d'autres cas, que ie ne m'areſte pas a deduire, à cauſe qu'ils n'ont eu aucun vſage en la Dioptrique.

On pourroit auſſy paſſer outre, & dire, lorſque l'vne des ſuperficies du verre eſt donnée, pourueû qu'elle ne ſoit que toute plate, ou compoſée de ſections coniques, ou de cercles; comment on doit faire ſon autre ſuperficie, affin qu'il tranſmette tous les rayons d'vn point donné, a vn autre point auſſy donné. car ce n'eſt rien de plus difficile que ceque ie viens d'expliquer; ou plutoſt c'eſt choſe beaucoup plus facile, à cauſe que le chemin en eſt ouuert. Mais i'ayme mieux, que d'autres le cherchent, affinque s'ils ont encore vn peu de peine à le trouuer, cela leur face d'autant plus eſtimer l'inuention des choſes qui ſont icy demonſtrées.

Commēt on peut appliquer ce qui a eſté dit icy des lignes courbes deſcrites ſur vne ſuperficie plate, à celles qui ſe deſcriuēt dās vn eſpace qui a trois dimenſions.

Au reſte ie n'ay parlé en tout cecy, que des lignes courbes, qu'on peut deſcrire ſur vne ſuperficie plate; mais il eſt ayſé de rapporter ceque i'en ay dit, à toutes celles qu'on ſçauroit imaginer eſtre formées, par le mouuement regulier des poins de quelque cors, dans vn eſpace qui a trois dimenſions. A ſçauoir en tirant deux perpendiculaires, de chaſcun des poins de la ligne courbe qu'on veut conſiderer, ſur deux plans qui s'entrecouppent a angles droits, l'vne ſur l'vn, & l'autre ſur l'autre. car les extremités de ces perpendiculaires deſcriuent deux autres lignes courbes, vne ſur chaſcun de ces plans, deſquelles on peut, en la façon cy deſſus expliquée, determiner tous

les

of one of the third class. Finally, if AM is equal to $\dfrac{ge}{d-e}$, the curves AC and CY must both be hyperbolas.

These two problems can be extended to an infinity of other cases which I will not stop to deduce, since they have no practical value in dioptrics.

I might go farther and show how, if one surface of a lens is given and is neither entirely plane nor composed of conic sections or circles, the other surface can be so determined as to transmit all the rays from a given point to another point, also given. This is no more difficult than the problems I have just explained; indeed, it is much easier since the way is now open; I prefer, however, to leave this for others to work out, to the end that they may appreciate the more highly the discovery of those things here demonstrated, through having themselves to meet some difficulties.

In all this discussion I have considered only curves that can be described upon a plane surface, but my remarks can easily be made to apply to all those curves which can be conceived of as generated by the regular movement of the points of a body in three-dimensional space.[178] This can be done by dropping perpendiculars from each point of the curve under consideration upon two planes intersecting at right angles, for the ends of these perpendiculars will describe two other curves, one in each of the two planes, all points of which may be determined in the way already explained, and all of which may be related to those of a straight line common to the two planes; and by means of these the points of the three-dimensional curve will be entirely determined.

[178] This is the hint which Descartes gives of the possibility of the extension of his theory to solid geometry. This extension was effected largely by Parent (1666-1716), Clairaut (1713-1765), and Van Schooten (d. 1661).

We can even draw a straight line at right angles to this curve at a given point, simply by drawing a straight line in each plane normal to the curve lying in that plane at the foot of the perpendicular drawn from the given point of the three-dimensional curve to that plane and then drawing two other planes, each passing through one of the straight lines and perpendicular to the plane containing it; the intersection of these two planes will be the required normal.

And so I think I have omitted nothing essential to an understanding of curved lines.

les poins, & les rapporter a ceux de la ligne droite , qui
eſt commune a ces deux plans, au moyen dequoy ceux
de la courbe, qui a trois dimenſions , ſont entierement
determinés. Meſme ſi on veut tirer vne ligne droite, qui
couppe cete courbe au point donné a angles droits · il
faut ſeulement tirer deux autres lignes droites dans les
deux plans, vne en chaſcun, qui couppent a angles droits
les deux lignes courbes, qui y ſont, aux deux poins , ou
tombent les perpendiculaires qui vienent de ce point
donné. car ayant eſleué deux autres plans , vn ſur chaſcune de ces lignes droites, qui couppe a angles droits le
plan où elle eſt, on aura l'interſection de ces deux plans
pour la ligne droite cherchée. Et ainſi ie penſe n'auoir
rien omis des elemens, qui ſont neceſſaires pour la connoiſſance des lignes courbes.

BOOK THIRD

Geometry

BOOK III

ON THE CONSTRUCTION OF SOLID AND SUPERSOLID PROBLEMS

WHILE it is true that every curve which can be described by a continuous motion should be recognized in geometry, this does not mean that we should use at random the first one that we meet in the construction of a given problem. We should always choose with

L A
GEOMETRIE.
LIVRE TROISIESME.

De la conſtruction des Problesmes , qui
ſont Solides, ou pluſque Solides.

De quel-
les lignes
courbes
on peut
ſe ſeruir,
en la con-
ſtruction
de chaſq;
probleſ-
me.

ENCORE que toutes les lignes courbes, qui peuuent
eſtre deſcrites par quelque mouuement regulier,
doiuent eſtre receuës en la Geometrie , ce n'eſt pas a dire qu'il ſoit permis de ſe ſeruir indifferemment de la premiere qui ſe rencontre, pour la conſtruction de chaſque
pro-

A a a

problefme: mais il faut auoir foin de choifir toufiours la plus fimple, par laquelle il foit poffible de le refoudre. Et mefme il eft a remarquer, que par les plus fimples on ne doit pas feulement entendre celles, qui pcuuent le plus ayfement eftre defcrites, ny celles qui rendent la conftruction, ou la demonftration du Problefme propo-féplus facile, mais principalement celles, qui font du plus fimple genre, qui puiffe feruir a determiner la quan-tité qui eft cherchée.

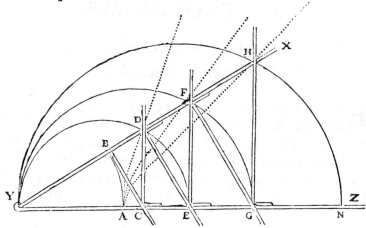

Exemple touchant l'inuentiõ de plu-fieurs moyēnes propro-tionelles. Comme par exemple ie ne croy pas, qu'il y ait aucu-ne façon plus facile, pour trouuer autant de moyennes proportionnelles, qu'on veut, ny dont la demonftration foit plus euidente, que d'y employer les lignes courbes, qui fe defcriuent par l'inftrument X Y Z cy deffus expli-qué. Car voulant trouuer deux moyennes proportion-nelles entre Y A & Y E, il ne faut que defcrire vn cercle, dont le diametre foit Y E; & pource que ce cercle couppe

care the simplest curve that can be used in the solution of a problem, but it should be noted that the simplest means not merely the one most easily described, nor the one that leads to the easiest demonstration or construction of the problem, but rather the one of the simplest class that can be used to determine the required quantity.

For example, there is, I believe, no easier method of finding any number of mean proportionals,[179] nor one whose demonstration is clearer, than the one which employs the curves described by the instrument XYZ, previously explained.[180]   Thus, if two mean proportionals between YA and YE be required, it is only necessary to describe

[179] For the history of this problem, see Heath, *History,* Vol. I, p. 244, et seq.
[180] See page 46.

a circle upon YE as diameter cutting the curve AD in D, and YD is then one of the required mean proportionals. The demonstration becomes obvious as soon as the instrument is applied to YD, since YA (or YB) is to YC as YC is to YD as YD is to YE.

Similarly, to find four mean proportionals between YA and YG, or six between YA and YN, it is only necessary to draw the circle YFG, which determines by its intersection with AF the line YF, one of the four mean proportionals; or the circle YHN, which determines by its intersection with AH the line YH, one of the six mean proportionals, and so on.

But the curve AD is of the second class, while it is possible to find two mean proportionals by the use of the conic sections, which are curves of the first class.[181] Again, four or six mean proportionals can be found by curves of lower classes than AF and AH respectively. It would therefore be a geometric error to use these curves. On the other hand, it would be a blunder to try vainly to construct a problem by means of a class of lines simpler than its nature allows.[182]

Before giving the rules for the avoidance of both these errors, some general statements must be made concerning the nature of equations. An equation consists of several terms, some known and some unknown, some of which are together equal to the rest; or rather, all of which taken together are equal to nothing; for this is often the best form to consider.[183]

[181] If we let x and y represent the two mean proportionals between a and b we have $a : x = x : y = y : b$, whence $x^2 = ay$; $y^2 = bx$, and $xy = ab$. Therefore x and y may be found by determining the intersections of two parabolas or of a parabola and a hyperbola.

[182] Cf. Pappus, Book IV, Prop. 31, Vol. I, p. 273. See also Guisnée, *Application de l'Algèbre a la Géométrie,* Paris, 1733, p. 28, and L'Hospital, *Traité Analytique des Sections Coniques,* Paris, 1707, p. 400.

[183] The advantage of this arrangement had been recognized by several writers before Descartes.

156

pe la courbe A D au point D, Y D eſt l'vne des moyennes
proportionnelles cherchées.   Dont la demonſtration ſe
voit a l'œil par la ſeule application de cet inſtrument ſur
la ligne Y D. car comme Y A, ou Y B, qui luy eſt eſgale
eſt a Y C; ainſi Y C eſt a Y D; & Y D a Y E.

Toutdemeſme pour trouuer quatre moyennes proportionelles entre Y A & Y G; ou pour en trouuer ſix entre Y A & Y N, il ne faut que tracer le cercle Y F G, qui
couppant A F au point F, determine la ligne droite Y F,
qui eſt l'vne de ces quatre proportionnelles; ou Y H N,
qui couppant A H au point H, determine Y H l'vne des
ſix, & ainſi des autres.

Mais pourceque la ligne courbe A D eſt du ſecond
genre, & qu'on peut trouuer deux moyenes proportionelles par les ſections coniques, qui ſont du premier ; &
auſſy pourcequ'on peut trouuer quatre ou ſix moyenes
proportionelles, par des lignes qui ne ſont pas de genres
ſi compoſés, que ſont A F, & A H, ce ſeroit vne faute en
Geometrie que de les y employer. Et c'eſt vne faute
auſſy d'autre coſté de ſe trauailler inutilement a vouloir
conſtruire quelque problefme par vn genre de lignes
plus ſimple, que ſa nature ne permet.

Or affin que ie puiſſe icy donner quelques reigles, *De la na-*
pour euiter l'vne & l'autre de ces deux fautes, il faut que *ture des*
ie die quelque choſe en general de la nature des Equa- *Equatiós.*
tions; c'eſt a dire des ſommes compoſées de pluſieurs termes partie connus, & partie inconnus, dont les vns ſont
eſgaux aux autres, ou plutoſt qui conſiderés tous enſemble ſont eſgaux a rien. car ce ſera ſouuent le meilleur de
les conſiderer en cete ſorte.

A a a 2   Scachés

LA GEOMETRIE.

Combien il peut y auoir de racines en chafq; Equatió.

Scachés donc qu'en chafque Equation, autant que la quantité inconnue a de dimenfions, autant peut il y auoir de diuerfes racines, c'eft a dire de valeurs de cete quantité. car par exemple fi on fuppofe x efgale a 2; oubien $x -- 2$ efgal a rien ; & derechef $x \infty 3$; oubien $x -- 3 \infty 0$; en multipliant ces deux equatiòns $x -- 2 \infty 0$, & $x -- 3 \infty 0$, l'vne par l'autre, on aura $xx -- 5x + 6 \infty 0$, oubien $xx \infty 5x -- 6$, qui eft vne Equation en laquelle la quantité x vaut 2 & tout enfemble vaut 3. Que fi derechef on fait $x -- 4 \infty 0$, & qu'on multiplie cete fomme par $xx -- 5x + 6 \infty 0$, on aura $x^3 -- 9xx + 26x -- 24 \infty 0$, qui eft vne autre Equation en laquelle x ayant trois dimenfions a auffy trois valeurs, qui font 2, 3, & 4.

Quelles font les fauffes racines.

Mais fouuent il arriue, que quelques vnes de ces racines font fauffes, ou moindres que rien. comme fi on fuppofe que x defigne auffy le defaut d'vne quantité, qui foit 5, on a $x + 5 \infty 0$, qui eftant multipliée par $x^3 -- 9xx + 26x -- 24 \infty 0$ fait

$$x^4 -- 4x^3 -- 19xx + 106x -- 120 \infty 0$$

pour vne equation en laquelle il y a quatre racines, a fçauoir trois vrayes qui font 2, 3, 4, & vne fauffe qui eft 5.

Côment on peut diminuer le nombre des dimenfions d'vne E-quation lorfqu'on connoift quel-qu'vne de fes racines.

Et on voit euidemment de cecy, que la fomme d'vne equation, qui contient plufieurs racines, peut toufiours eftre diuifée par vn binóme compofé de la quantité inconnuë, moins la valeur de l'vne des vrayes racines, laquelle que ce foit; ou plus la valeur de l'vne des fauffés. Au moyen de quoy on diminue d'autant fes dimenfions.

Et reciproquement que fi la fomme d'vne equation
ne

Every equation can have[184] as many distinct roots (values of the unknown quantity) as the number of dimensions of the unknown quantity in the equation.[185] Suppose, for example, $x = 2$ or $x-2 = 0$, and again, $x = 3$, or $x-3 = 0$. Multiplying together the two equations $x-2 = 0$ and $x-3 = 0$, we have $x^2-5x+6 = 0$, or $x^2 = 5x-6$. This is an equation in which x has the value 2 and at the same time[186] x has the value 3. If we next make $x-4 = 0$ and multiply this by $x^2-5x+6 = 0$, we have $x^3-9x^2+26x-24 = 0$ another equation, in which x, having three dimensions, has also three values, namely, 2, 3, and 4.

It often happens, however, that some of the roots are false[187] or less than nothing. Thus, if we suppose x to represent the defect[188] of a quantity 5, we have $x+5 = 0$ which, multiplied by $x^3-9x^2+26x-24 = 0$, yields $x^4-4x^3-19x^2+106x-120 = 0$, an equation having four roots, namely three true roots, 2, 3, and 4, and one false root, 5.[189]

It is evident from the above that the sum[190] of an equation having several roots is always divisible by a binomial consisting of the unknown quantity diminished by the value of one of the true roots, or plus the value of one of the false roots. In this way,[191] the degree of an equation can be lowered.

On the other hand, if the sum of the terms of an equation[192] is not divisible by a binomial consisting of the unknown quantity plus or

[184] It is worthy of note that Descartes writes "can have" ("peut-il y avoir"), not "must have," since he is considering only real positive roots.

[185] That is, as the number denoting the degree of the equation.

[186] "Tout ensemble,"—not quite the modern idea.

[187] "Racines fausses," a term formerly used for "negative roots." Fibonacci, for example, does not admit negative quantities as roots of an equation. *Scritti de Leonardo Pisano,* published by Boncompagni, Rome, 1857. Cardan recognizes them, but calls them "æstimationes falsæ" or "fictæ," and attaches no special significance to them. See Cardan, *Ars Magna,* Nurnberg, 1545, p. 2. Stifel called them "Numeri absurdi," as also in Rudolff's Coss, 1545.

[188] "Le défaut." If $x = -5$, -5 is the "defect" of 5, that is, the remainder when 5 is subtracted from zero.

[189] That is, three positive roots, 2, 3, and 4, and one negative root, -5.

[190] "Somme," the left member when the right member is zero; that is, what we represent by $f(x)$ in the equation $f(x)=0$.

[191] That is, by performing the division.

[192] "Si la somme d'un équation."

minus some other quantity, then this latter quantity is not a root of the equation. Thus the[103] above equation $x^4 - 4x^3 - 19x^2 + 106x - 120 = 0$ is divisible by $x-2$, $x-3$, $x-4$ and $x+5$,[104] but is not divisible by x plus or minus any other quantity. Therefore the equation can have only the four roots, 2, 3, 4, and 5.[105] We can determine also the number of true and false roots that any equation can have, as follows:[106] An equation can have as many true roots as it contains changes of sign, from $+$ to $-$ or from $-$ to $+$; and as many false roots as the number of times two $+$ signs or two $-$ signs are found in succession.

Thus, in the last equation, since $+x^4$ is followed by $-4x^3$, giving a change of sign from $+$ to $-$, and $-19x^2$ is followed by $+106x$ and $+106x$ by -120, giving two more changes, we know there are three true roots; and since $-4x^3$ is followed by $-19x^2$ there is one false root.

It is also easy to transform an equation so that all the roots that were false shall become true roots, and all those that were true shall become false. This is done by changing the signs of the second, fourth,

---

[103] First member of the equation. Descartes always speaks of dividing the equation.

[104] Incorrectly given as $x-5$ in some editions.

[105] Where 5 would now be written -5. Descartes neither states nor explicitly assumes the fundamental theorem of algebra, namely, that every equation has at least one root.

[106] This is the well known "Descartes's Rule of Signs." It was known however, before his time, for Harriot had given it in his *Artis analyticae praxis*, London, 1631. Cantor says Descartes may have learned it from Cardan's writings, but was the first to state it as a general rule. See Cantor, Vol. II(1) pp. 496 and 725.

Cóment
on peut
examiner
ſi quelque
quantité
donnée
eſt la va-
leur d'vne
racine.

ne peut eſtre diuiſée par vn binóme compoſé de la quan-
titéinconnue ✛ ou -- quelque autre quantité, cela teſmoigne que cete autre quantité n'eſt la valeur d'aucune
de ſes racines. Comme cete derniere

$$x^4 -- 4x^3 -- 19xx ✛ 106x -- 120 \infty 0$$

peut bien eſtre diuiſée, par x -- 2, & par x -- 3, & par
x -- 4, & par x ✛ 5; mais non point par x ✛ ou -- aucune autre quantité. cequi monſtre qu'elle ne peut auoir
que les quatre racines 2,3,4,& 5.

Combien
il peut y
auoir de
vrayes
racines en
chaſque
Equatió.

On connoiſt auſſy de cecy combien il peut y auoir de
vrayes racines, & combien de fauſſes en chaſque Equation. A ſçauoir il y en peut auoir autant de vrayes, que
les ſignes ✛ & -- s'y trouuent de fois eſtre changés ; &
autant de fauſſes qu'il s'y trouue de fois deux ſignes ✛,
ou deux ſignes -- qui s'entreſuiuent. Comme en la derniere, a cauſe qu'aprés ✛ x^4 il y a -- $4x^3$, qui eſt vn chan-
gement du ſigne ✛ en --, & aprés -- 19 xx il y a ✛ 106x,
& aprés ✛ 106 x il y a -- 120 qui ſont encore deux autres
changemens, on connoiſt qu'il y a trois vrayes racines; &
vne fauſſe, a cauſe que les deux ſignes --, de 4 x^3, & 19 xx,
s'entreſuiuent.

Cóment
on fait
que les
fauſſes
racines
d'vne E-
quation
deuienét
vrayes, &
les vrayes
fauſſes.

De plus il eſt ayſé de faire en vne meſme Equation,
que toutes les racines qui eſtoient fauſſes deuienent
vrayes, & par meſme moyen que toutes celles qui eſtoiét
vrayes deuienent fauſſes : a ſçauoir en changeant tous
les ſignes ✛ ou -- qui ſont en la ſeconde, en la
quatrieſme, en la ſixieſme, ou autres places qui ſe
deſignent par les nombres pairs, ſans changer ceux
de la premiere, de la troiſieſme, de la cinquieſme
& ſemblables qui ſe deſignent par les nombres
Aaa 3 impairs.

impairs. Comme si au lieu de

$$+x^4 -- 4x^3 -- 19xx + 106x -- 120 \infty 0$$

on escrit

$$+x^4 + 4x^3 -- 19xx -- 106x -- 120 \infty 0$$

on a vne Equation en laquelle il n'y a qu'vne vraye racine, qui est 5, & trois fausses qui sont 2, 3, & 4.

Que si sans connoistre la valeur des racines d'vne E- quation, on la veut augmenter, ou diminuer de quelque quantité connuë, il ne faut qu'au lieu du terme inconnu en supposer vn autre, qui soit plus ou moins grand de cete mesme quantité, & le substituer par tout en la place du premier.

Comme si on veut augmenter de 3 la racine de cete Equation

$$x^4 + 4x^3 -- 19xx -- 106x -- 120 \infty 0$$

il faut prendre y au lieu d'x, & penser que cete quantité y est plus grande qu'x de 3, en sorte que $y -- 3$ est esgal a x, & au lieu d'xx, il faut mettre le quarré d'$y -- 3$ qui est $yy -- 6y + 9$ & au lieu d'x^3 il faut mettre son cube qui est $y^3 -- 9yy + 27y -- 27$, & enfin au lieu d'x^4 il faut mettre son quarré de quarré qui est $y^4 -- 12y^3 + 54yy -- 108y + 81$. Et ainsi descriuant la somme precedente en substituant par tout y au lieu d'x on a

$$y^4 -- 12y^3 + 54yy -- 108y + 81$$
$$+ 4y^3 -- 36yy + 108y -- 108$$
$$-- 19yy + 114y -- 171$$
$$-- 106y + 318$$
$$-- 120$$

$$\overline{y^4 -- 8y^3 -- 1yy \quad + 8y^* \qquad \infty 0}$$

oubien

sixth, and all even terms, leaving unchanged the signs of the first, third, fifth, and other odd terms. Thus, if instead of

$$+x^4-4x^3-19x^2+106x-120 = 0$$

we write

$$+x^4+4x^3-19x^2-106x-120 = 0$$

we get an equation having one true root, 5, and three false roots, 2, 3, and 4.[107]

If the roots of an equation are unknown and it be desired to increase or diminish each of these roots by some known number, we must substitute for the unknown quantity throughout the equation, another quantity greater or less by the given number. Thus, if it be desired to increase by 3 the value of each root of the equation

$$x^4+4x^3-19x^2-106x-120 = 0$$

put y in the place of x, and let y exceed x by 3, so that $y-3 = x$. Then for x^2 put the square of $y-3$, or y^2-6y+9; for x^3 put its cube, $y^3-9y^2+27y-27$; and for x^4 put its fourth power,[108] or

$$y^4-12y^3+54y^2-108y+81.$$

Substituting these values in the above equation, and combining, we have

$$
\begin{array}{r}
y^4 - 12y^3 + 54y^2 - 108y + 81 \\
+ 4y^3 - 36y^2 + 108y - 108 \\
- 19y^2 + 114y - 171 \\
- 106y + 318 \\
- 120 \\
\hline
y^4 - 8y^3 - y^2 + 8y = 0,^{[109]}
\end{array}
$$

or

$$y^3-8y^2-y+8 = 0,$$

[107] In absolute value.

[108] "Son quarré de quarré," that is, its fourth power.

[109] Descartes wrote this $y^4 - 8y^3 - y^2 + 8y \ast \infty\ 0$, indicating by a star the absence of a term in a complete polynomial.

whose true root is now 8 instead of 5, since it has been increased by 3. If, on the other hand, it is desired to diminish by 3 the roots of the same equation, we must put $y+3 = x$ and $y^2+6y+9 = x^2$, and so on. so that instead of $x^4 + 4x^3 - 19x^2 - 106x - 120 = 0$, we have

$$
\begin{array}{l}
y^4 + 12y^3 + 54y^2 + 108y + 81 \\
 + 4y^3 + 36y^2 + 108y + 108 \\
 - 19y^2 - 114y - 171 \\
 - 106y - 318 \\
 - 120 \\
\hline
y^4 + 16y^3 + 71y^2 - 4y - 420 = 0.
\end{array}
$$

It should be observed that increasing the true roots of an equation diminishes[200] the false roots by the same amount; and on the contrary diminishing the true roots increases the false roots; while diminishing either a true or a false root by a quantity equal to it makes the root zero; and diminishing it by a quantity greater than the root renders a true root false or a false root true.[201] Thus by increasing the true root 5 by 3, we diminish each of the false roots, so that the root previously 4 is now only 1, the root previously 3 is zero, and the root previously 2 is now a true root, equal to 1, since $-2+3 = +1$. This explains why the equation $y^3-8y^2-y+8 = 0$ has only three roots,

[200] In absolute value.

[201] For example, the false root 5 diminished by 7 means $-(5-7) = +2$.

oubien $y^3 -- 8yy -- 1y + 8 \infty 0$.

où la vraye racine qui eſtoit 5 eſt maintenant 8 , a cauſe du nombre trois qui luy eſt aiouſté.

Que ſi on veut au contraire diminuer de trois la racine de cete meſme Equation , il faut faire $y + 3 \infty x$ & $yy + 6y + 9 \infty xx$. & ainſi des autres de façon qu'au lieu de

$$x^4 + 4x^3 -- 19xx -- 106x -- 120 \infty 0$$

on met

$$y_4 + 12y^3 + 54yy + 108y + 81$$
$$+ 4y^3 + 36 \ yy + 108 \ y + 108$$
$$-- 19 \ yy \ -- \ 114 \ y \ -- \ 171$$
$$-- \ 106y \ -- \ 318$$
$$-- 120$$

$$y^4 + 16y^3 + 71yy -- \ \ 4y \ -- 420 \infty 0.$$

Et il eſt a remarquer qu'en augmentant les vrayes racines d'vne Equation, on diminue les fauſſes de la meſme quantité; ou au contraire en diminuant les vrayes,on augmente les fauſſes. Et que ſi on diminue ſoit les vnes ſoit les autres, d'vne quantité qui leur ſoit eſgale, elles deuienent nulles, & que ſi c'eſt d'vne quantité qui les ſurpaſſe, de vrayes elles deuienent fauſſes, ou de fauſſes vrayes. Comme icy en augmentant de 3 la vraye racine qui eſtoit 5, on a diminué de 3 chaſcune des fauſſes , en ſorte que celle qui eſtoit 4 n'eſt plus qu' 1, & celle qui eſtoit 3 eſt nulle, & celle qui eſtoit 2 eſt deuenue vraye & eſt 1, a cauſe que -- 2 + 3 fait + 1. c'eſt pourquoy en cete Equation $y^3 -- 8yy -- 1y + 8 \infty 0$ il ny a plus que 3 racines, entre leſquelles il y en a deux qui ſont vrayes,

I. &

Qu'en augmentant les vrayes racines on diminue les fauſſes, & au contraire.

1, & 8, & vne fauſſe qui eſt auſſy 1. & en cete autre

$$y^4 + 16y^3 + 71yy - 4y - 420 \infty 0$$

il n'y en a qu'vne vraye qui eſt 2, a cauſe que $+5-3$ fait $+2$, & trois fauſſes qui ſont 5, 6, & 7.

Cõment on peut oſter le ſecond terme d'vne E-quation.

Or par cete façon de changer la valeur des racines ſans les connoiſtre, on peut faire deux choſes, qui auront cy aprés quelque vſage: la premiere eſt qu'on peut touſiours oſter le ſecond terme de l'Equation qu'on examine, a ſçauoir en diminuant les vrayes racines, de la quantité connuë de ce ſecond terme diuiſée par le nombre des dimenſions du premier, ſi l'vn de ces deux termes eſtant marqué du ſigne $+$, l'autre eſt marqué du ſigne $-$; oubien en l'augmentant de la meſme quantité, s'ils ont tous deux le ſigne $+$, ou tous deux le ſigne $-$. Comme pour oſter le ſecond terme de la derniere Equatiõ qui eſt

$$y^4 + 16y^3 + 71yy - 4y - 420 \infty 0$$

ayant diuiſé 16 par 4, a cauſe des 4 dimenſions du terme y_4, il vient derechef 4, c'eſt pourquoy ie fais $z - 4 \infty y$, & i'eſcris

$$z^4 - 16z^3 + 96zz - 256z + 256$$
$$+ 16z^3 - 192zz + 768z - 1024$$
$$+ 71zz - 568z + 1136$$
$$- 4z + 16$$
$$- 420$$

$$z^4 \quad * \quad -25zz - 60 \quad z - 36 \infty 0.$$

ou la vraye racine qui eſtoit 2, eſt 6, a cauſe qu'elle eſt augmentée de 4; & les fauſſes qui eſtoient 5, 6, & 7, ne ſont plus que 1, 2, & 3, a cauſe qu'elles ſont diminuées chaſcune de 4.

　　　　　　　　　　　　　　　　　　Tout

two of them, 1 and 8, being true roots, and the third, also 1, being false; while the other equation $y^4-16y^3+71y^2-4y-420=0$ has only one true root, 2, since $+5-3=+2$, and three false roots, 5, 6, and 7.

Now this method of transforming the roots of an equation without determining their values yields two results which will prove useful: First, we can always remove the second term of an equation by diminishing its true roots by the known quantity of the second term divided by the number of dimensions of the first term, if these two terms have opposite signs; or, if they have like signs, by increasing the roots by the same quantity.[202] Thus, to remove the second term of the equation $y^4+16y^3+71y^2-4y-420=0$ I divide 16 by 4 (the exponent of y in y^4), the quotient being 4. I then make $z-4=y$ and write

$$
\begin{array}{r}
z^4 - 16z^3 + 96z^2 - 256z + 256 \\
+ 16z^3 - 192z^2 + 768z - 1024 \\
+ 71z^2 - 568z + 1136 \\
- 4z + 16 \\
- 420 \\
\hline
z^4 \qquad - 25z^2 - 60z - 36 = 0.
\end{array}
$$

The true root of this equation which was 2 is now 6, since it has been increased by 4, and the false roots, 5, 6, and 7, are only 1, 2, and 3,

[202] That is, by diminishing the roots by a quantity equal to the coefficient of the second term divided by the exponent of the highest power of x, with the opposite sign.

167

since each has been diminished by 4. Similarly, to remove the second terms of $x^4-2ax^3+(2a^2-c^2)x^2-2a^3x+a^4=0$; since $2a\div 4=\dfrac{1}{2}a$ we must put $z+\dfrac{1}{2}a=x$ and write

$$z^4 + 2az^3 + \frac{3}{2}a^2z^2 + \frac{1}{2}a^3z + \frac{1}{16}a^4$$
$$- 2az^3 - 3a^2z^2 - \frac{3}{2}a^3z - \frac{1}{4}a^4$$
$$+ 2a^2z^2 + 2a^3z + \frac{1}{2}a^4$$
$$- c^2z^2 - ac^2z - \frac{1}{4}a^2c^2$$
$$- 2a^3z - a^4$$
$$+ a^4$$
$$\overline{z^4 + \left(\frac{1}{2}a^2-c^2\right)z^2 - (a^3+ac^2)z + \frac{5}{16}a^4 - \frac{1}{4}a^2c^2 = 0.}$$

Having found the value of z, that of x is found by adding $\dfrac{1}{2}a$. Second, by increasing the roots by a quantity greater than any of the false roots[203] we make all the roots true. When this is done, there will be no two consecutive $+$ or $-$ terms; and further, the known quantity of the third term will be greater than the square of half that of the second term. This can be done even when the false roots are unknown, since approximate values can always be obtained for them and the roots can then be increased by a quantity as large as or larger than is required. Thus, given,

[203] In absolute value.

Tout de mefme fi on veut ofter le fecond terme de

$$x^4 - 2a x^3 \overset{+\ 2aa}{\underset{-\ cc}{}} xx - 2a^3 x + a^4 \infty\, 0,$$

pourceque diuifant $2\,a$ par 4 il vient $\frac{1}{2}\,a$; il faut faire
$\zeta + \frac{1}{2}\,a \infty\, x$.& efcrire

$$\zeta^4 + 2a\zeta^3 + \tfrac{3}{2}aa\zeta\zeta + \tfrac{1}{2}a^3\zeta + \tfrac{1}{16}a^4$$
$$- 2a\zeta^3 \quad - 3aa\zeta\zeta - \tfrac{2}{2}a^3\zeta \quad - \tfrac{1}{4}a^4$$
$$+ 2aa\zeta\zeta + 2a^3 \quad + \tfrac{1}{2}a^4$$
$$- cc \quad - acc \quad - \tfrac{1}{4}aacc$$
$$- 2a^3 \quad - a^4$$
$$+ a^4$$

$$\zeta^4 \quad * \quad + \tfrac{1}{2}aa\zeta\zeta - a^3\ \zeta + \tfrac{5}{16}a^4 \ \infty\, 0$$
$$- cc \quad - acc \quad - \tfrac{3}{4}aacc$$

& fi on trouue aprés la valeur de ζ, en luy adiouftant $\frac{1}{2}\,a$
on aura celle de x.

La feconde chofe, qui aura cy aprés quelque vfage, Cóment
on peut
faire que
toutés
les fauffes
racines
d'vne
Equation
deuienēt
vrayes,
fans que
les vrayes
deuienēt
fauffes. eft, qu'on peut toufiours en augmentant la valeur des
vrayes racines, d'vne quantité qui foit plus grande que
n'eft celle d'aucune des fauffos, faire qu'elles deuienent
toutes vrayes, en forte qu'il n'y ait point deux fignes +,
ou deux fignes -- qui s'entrefuiuent, & outre cela que la
quantité connuë du troifiefme terme foit plus grande,
que le quarré de la moitié de celle du fecond. Car encore que cela fe face, lorfque ces fauffes racines font
inconnuës, il eft ayfé neanmoins de iuger a peu pré. de
leur grandeur, & de prendre vne quantité, qui les fur-
paffe d'autant, ou de plus, qu'il n'eft requis a cet effect.
Comme fi on a

Bbb

$$x^6 + nx^5 - 6nnx^4 + 36n^3x^3 - 216n^4x^2 + 1296n^5x - 7776n^6 \infty 0.$$

en faifant $y - 6n \infty x$, on trouuera

$$
\begin{array}{l}
y^6 - 36n \; y^5 + 540nn \; y^4 - 4320n^3 \; y^3 + 19440n^4 \; yy - 46656n^5 \; y + 46656n^6 \\
\quad + n \quad\quad - 30nn \quad + 360n^3 \quad - 2160n^4 \quad + 6480n^5 \quad - 7776n^6 \\
\quad\quad - 6nn \quad + 144n^3 \quad - 1296n^4 \quad + 5184n^5 \quad - 7776n^6 \\
\quad\quad\quad + 36n^3 \quad - 648n^4 \quad + 3888n^5 \quad - 7776n^6 \\
\quad\quad\quad\quad - 216n^4 \quad + 2592n^5 \quad - 7776n^6 \\
\quad\quad\quad\quad\quad + 1296n^5 \quad - 7776n^6
\end{array}
$$

$$y^6 - 35n \; y^5 + 504nn \; y^4 - 3780n^3 \; y^3 + 15120n^4 \; y^2 - 27216n^5 y \; \infty 0.$$

Ou il .eſt manifeſte, que 504 nn, qui eſt la quantité connuë du troiſieſme terme eſt plus grande, que le quarré de $\frac{35}{2}n$, qui eſt la moitié de celle du ſecond. Et il n'y a point de cas, pour lequel la quantité, dont on augmente les vrayes racines, ait beſoin a cet effect, d'eſtre plus grande, a proportion de celles qui ſont données, que pour cetuy cy.

Cōment on fait que tou-tes les places d'vne E-quation ſoient remplies.

Mais a cauſe que le dernier terme s'y trouue nul, ſi on ne deſire pas que cela ſoit, il faut encore augmenter tant ſoit peu la valeur des racines ; Et ce ne ſçauroit eſtre de ſi peu, que ce ne ſoit aſſés pour cet effect. Non plus que lorſqu'on veut accroiſtre le nombre des dimenſions de quelque Equation, & faire que toutes les places de ſes termes ſoient remplies. Comme ſi au lieu de x^5 **** $- 6 \infty 0$, on veut auoir vne Equation, en laquelle la quantité inconnue ait ſix dimenſions, & dont aucun des termes ne ſoit nul, il faut premierement pour

$$x^5 \ast \; \ast \; \ast \; \ast - b \infty 0 \; \text{eſcrire}$$
$$x^6 \ast \; \ast \; \ast \; \ast - bx \; \ast \infty 0$$

puis ayant fait $y - a \infty x$, on aura

$$y^6 - 6ay^5 + 15aay^4 - 20a^3y^3 + 15a^4yy - 6a^5y + a^6$$
$$- by + ab \infty 0$$

Qu il eſt manifeſte que tant petite que la quantité a ſoit ſuppoſée

$$x^6+nx^5-6n^2x^4+36n^3x^3-216n^4x^2+1296n^5x-7776n^6 = 0,$$

make $y-6n = x$ and we have,

$$
\begin{array}{l}
y^6-36n \left.\begin{array}{l} y^5+540n^2 \\ +\ n \end{array}\right\} \ \left.\begin{array}{l} y^4-4320n^3 \\ -\ 30n^2 \\ -\ 6n^2 \end{array}\right\} \ \left.\begin{array}{l} y^3+19440n^4 \\ +\ 360n^3 \\ +\ 144n^3 \\ +\ 36n^3 \end{array}\right\} \ \left.\begin{array}{l} y^2-46656n^5 \\ -\ 2160n^4 \\ -\ 1296n^4 \\ -\ 648n^4 \\ -\ 216n^4 \end{array}\right\} \ \left.\begin{array}{l} y+46656n^6 \\ +\ 6480n^5 \\ +\ 5184n^5 \\ +\ 3888n^5 \\ +\ 2592n^5 \\ +\ 1296n^5 \end{array}\right\} \ \begin{array}{l} -\ 7776n^6 \\ -\ 7776n^6 \\ -\ 7776n^6 \\ -\ 7776n^6 \\ -\ 7776n^6 \\ -\ 7776n^6 \\ -\ 7776n^6 \end{array}
\end{array}
$$

$$y^6-35ny^5 +504n^2y^4 -3780n^3y^3 +15120n^4y^2 -27216n^5y = 0.$$

Now it is evident that $504n^2$, the known quantity[204] of the third term, is larger than $\left(\dfrac{35}{2}n\right)^2$; that is, than the square of half that of the second term; and there is no case for which the true roots need be increased by a quantity larger in proportion to those given than for this one.

If it is undesirable to have the last term zero, as in this case, the roots must be increased just a little more, yet not too little, for the purpose. Similarly if it is desired to raise the degree of an equation, and also to have all its terms present, as if instead of $x^5-b = 0$, we wish an equation of the sixth degree with no term zero, first, for $x^5 - b = 0$ write $x^6- bx = 0$, and letting $y - a = x$ we have

$$y^6-6ay^5+15a^2y^4-20a^3y^3+15a^4y^2-(6a^5+b)y+a^6+ab = 0.$$

It is evident that, however small the quantity a, every term of this equation must be present.

[204] I. e., the coefficient.

We can also multiply or divide all the roots of an equation by a given quantity, without first determining their values. To do this, suppose the unknown quantity when multiplied or divided by the given number to be equal to a second unknown quantity. Then multiply or divide the known quantity of the second term by the given quantity, that in the third term by the square of the given quantity, that in the fourth term by its cube, and so on, to the end.

This device is useful in changing fractional terms of an equation, to whole numbers, and often[205] in rationalizing the terms. Thus, given $x^3 - \sqrt{3}\,x^2 + \dfrac{26}{27}\,x - \dfrac{8}{27\sqrt{3}} = 0$, let there be required another equation in which all the terms are expressed in rational numbers. Let $y = \sqrt{3}$ and multiply the second term by $\sqrt{3}$, the third by 3, and the last by $3\sqrt{3}$. The resulting equation is $y^3 - 3y^2 + \dfrac{26}{9}\,y - \dfrac{8}{9} = 0$. Next let it be required to replace this equation by another in which the known quantities are expressed only by whole numbers. Let $z = 3y$. Multiplying 3 by 3, $\dfrac{26}{9}$ by 9, and $\dfrac{8}{9}$ by 27, we have

$$z^3 - 9z^2 + 26z - 24 = 0.$$

The roots of this equation are 2, 3, and 4; and hence the roots of the

[205] But not always. Compare the case mentioned on page 175.

ſuppoſee toutes les places de l'Equation ne laiſſent pas
d'eſtre remplies.

De plus on peut, ſans connoiſtre la valeùr des vrayes
racines d'vne Equation, les multiplier, ou diuiſer toutes, par telle quantité connuë qu on veut. Ce qui ſe fait
en ſuppoſant que la quantité inconnuë eſtant multipliée,
ou diuiſée, par celle qui doit multiplier, ou diuiſer les
racines, eſt eſgale a quelque autre. Puis multipliant, ou
diuiſant la quantité connuë du ſecond terme, par cete
meſme qui doit multiplier, ou diuiſer les racines; & par
ſon quarré, celle du troiſieſme; & par ſon cube, celle du
quatrieſme; & ainſi iuſques au dernier. Ce qui peut ſeruir pour reduire a des nombres entiers & rationaux, les
fractions, ou ſouuent auſſy les nombres ſours, qui ſe
trouuent dans les termes des Equations. Comme ſi on a

$$x^3 - \sqrt{3}\ xx + \tfrac{26}{27}x - \tfrac{8}{27\sqrt{3}} \infty\ 0,$$

& qu'on veuille en auoir vne autre en ſa place, dont tous
les termes s'expriment par des nombres rationaux; il faut
ſuppoſer $y \infty x\sqrt{3}$, & multiplier par $\sqrt{3}$ la quantité
connuë du ſecond terme, qui eſt auſſy $\sqrt{3}$, & par ſon
quarré qui eſt 3 celle du troiſieſme qui eſt $\tfrac{26}{27}$, & par ſon
cube qui eſt $3\sqrt{3}$ celle du dernier, qui eſt $\tfrac{8}{27\sqrt{3}}$, ce qui
fait

$$y^3 - 3yy + \tfrac{26}{9}y - \tfrac{8}{9} \infty\ 0$$

Puis ſi on en veut auoir encore vne autre en la place de
celle cy, dont les quantites connuës ne s'expriment que
par des nombres entiers; il faut ſuppoſer $z \infty 3y$, & multipliant 3 par 3, $\tfrac{26}{9}$ par 9, & $\tfrac{8}{9}$ par 27 on trouue

$$z^3 - 9zz + 26z - 24 \infty\ 0,$$ où les racines eſtant 2, 3,
& 4, on connoiſt de là que celles de l'autre d'auparauant

Bbb 2   eſtoient

(marginal notes:) Commēt on peut plier ou diuiſer les racines ſans les connoiſtre.

Cóment on reduiſt les nombres rompus d'vne Equation a des entiers.

eſtoient $\frac{2}{3}$, 1, & $\frac{4}{3}$, & que celles de la premiere eſtoient $\frac{2}{9}\sqrt{3}$, $\frac{1}{3}\sqrt{3}$, & $\frac{4}{9}\sqrt{3}$.

Cóment on rend la quantité connuë de l'vn des termes d'vne Equation eſgale a telle autre qu'on veut. Cete operation peut auſſy ſeruir pour rendre la quantité connuë de quelqu'un des termes de l'Equatiõ eſgale a quelque autre donnée, comme ſi ayant

$$x^3 \quad * \quad -- bbx + c^3 \infty 0.$$

On veut auoir en ſa place vne autre Equation, en laquelle la quantité connuë, du terme qui occupe là troiſieſme place, a ſçauoir celle qui eſt icy bb, ſoit $3aa$, il faut ſuppoſer $y \infty x \sqrt{\frac{3aa}{bb}}$; puis eſcrire $y^3 \quad * \quad -- 3aay + \frac{3a^3c^3}{b^3}\sqrt{3} \infty 0$.

Que les racines, tant vrayes que fauſſes peuuent eſtre reelles ou imaginaires. Au reſte tant les vrayes racines que les fauſſes ne ſont pas touſiours reelles; mais quelquefois ſeulement imaginaires; c'eſt a dire qu'on peut bien touſiours en imaginer autant que iay dit en chaſque Equation; mais qu'il n'y a quelquefois aucune quantité, qui correſponde a celles qu'on imagine. comme encore qu'on en puiſſe imaginer trois en celle cy, $x^3 -- 6xx + 13x -- 10 \infty 0$, il n'y en a toutefois qu'vne reelle, qui eſt 2, & pour les deux autres, quoy qu'on les augmente, ou diminue, ou multiplie en la façon que ie viens d'expliquer, on ne ſçauroit les rendre autres qu'imaginaires.

La réduction des Equatiõs cubiques lorſque le probleſme eſt plan. Or quand pour trouuer la conſtruction de quelque probleſme, on vient a vne Equation, en laquelle la quantité inconnuë a trois dimenſions; premierement ſi les quantités connuës, qui y ſont, contienent quelques nombres rompus, il les faut reduire a d'autres entiers, par la multiplication tantoſt expliquée; Et s'ils en contienent de ſours, il faut auſſy les reduire a d'autres rationaux, autant qu'il ſera poſſible, tant par cete meſme multiplication,

preceding equation are $\frac{2}{3}$, 1 and $\frac{4}{3}$, and those of the first equation are

$$\frac{2}{9}\sqrt{3}, \frac{1}{3}\sqrt{3}, \text{ and } \frac{4}{9}\sqrt{3}.$$

This method can also be used to make the known quantity of any term equal to a given quantity. Thus, given the equation

$$x^3 - b^2x + c^3 = 0,$$

let it be required to write an equation in which the coefficient of the third term,[206] namely b^2, shall be replaced by $3a^2$. Let

and we have

$$y = x\sqrt{\frac{3a^2}{b^2}}$$

$$y^3 - 3a^2y + \frac{3a^3c^3}{b^3}\sqrt{3} = 0.$$

Neither the true nor the false roots are always real; sometimes they are imaginary;[207] that is, while we can always conceive of as many roots for each equation as I have already assigned,[208] yet there is not always a definite quantity corresponding to each root so conceived of. Thus, while we may conceive of the equation $x^3 - 6x^2 + 13x - 10 = 0$ as having three roots, yet there is only one real root, 2, while the other two, however we may increase, diminish, or multiply them in accordance with the rules just laid down, remain always imaginary.

When the construction of a problem involves the solution of an equation in which the unknown quantity has three dimensions,[209] the following steps must be taken:

First, if the equation contains some fractional coefficients,[210] change them to whole numbers by the method explained above;[211] if it con-

[206] Descartes wrote this equation $x \quad * \quad - bbx + c^3 \propto 0$, the star showing, as explained on page 163, that a term is missing. Hence, he speaks of $-b^2x$ as the third term.

[207] "Mais quelquefois seulement imaginaires." This is a rather interesting classification, signifying that we may have positive and negative roots that are imaginary. The use of the word "imaginary" in this sense begins here.

[208] This seems to indicate that Descartes realized the fact that an equation of the nth degree has exactly n roots. Cf. Cantor, Vol. II(1), p. 724.

[209] That is, a cubic equation.

[210] "Nombres rompues," the "numeri fracti" of the medieval Latin writers and "numeri rotti" of the Italians. The expression "broken numbers" was often used by early English writers.

[211] That is, transform the equation into one having integral coefficients.

tains surds, change them as far as possible into rational numbers, either by multiplication or by one of several other methods easy enough to find. Second, by examining in order all the factors of the last term, determine whether the left member of the equation is divisible[212] by a binomial consisting of the unknown quantity plus or minus any one of these factors. If it is, the problem is plane, that is, it can be constructed by means of the ruler and compasses; for either the known quantity of the binomial is the required root[213] or else, having divided the left member of the equation by the binomial, the quotient is of the second degree, and from this quotient the root can be found as explained in the first book.[214]

Given, for example, $y^6 - 8y^4 - 124y^2 - 64 = 0$.[215] The last term, 64, is divisible by 1, 2, 4, 8, 16, 32, and 64; therefore we must find whether the left member is divisible by $y^2 - 1$, $y^2 + 1$, $y^2 - 2$, $y^2 + 2$, $y^2 - 4$, and so on. We shall find that it is divisible by $y^2 - 16$ as follows:

$$
\begin{array}{l}
+ y^6 - 8y^4 - 124y^2 - 64 = 0 \\
\underline{- y^6 - 8y^4 - 4y^2} \\
 0 - 16y^4 - 128y^2 \quad -16 \\
\underline{- 16 - 16} \\
 + y^4 + 8y^2 + 4 = 0
\end{array}
$$

Beginning with the last term, I divide -64 by -16 which gives $+4$; write this in the quotient; multiply $+4$ by $+y^2$ which gives $+4y^2$ and

[212] "Qui divise toute la somme."
[213] That is, the root that satisfies the conditions of the problem.
[214] See page 13.
[215] Descartes considers this equation as a function of y^2.

tiplication, que par diuers autres moyens, qui font affés faciles a trouuer. Puis examinant par ordre toutes les quantités, qui peuuent diuiſer ſans fraction le dernier terme, il faut voir, ſi quelqu'vne d'elles, iointe auec la quantité inconnué par le ſigne $+$ ou $-$, peut compoſer vn binome, qui diuiſe toute la ſomme; & ſi cela eſt le Problefme eſt plan, c'eſt a dire il peut eſtre conſtruit auec la reigle & de compas; Car oubien la quantité connuë de ce binoſme eſt la racine cherchée; oubien l'Equation eſtant diuiſée par luy, ſe reduiſt a deux dimenſions, en ſorte qu'on en peut trouuer aprés la racine, par ce qui a eſté dit au premier liure.

Par exemple ſi on a

$$y^6 - 8y^4 - 124y^2 - 64 \infty 0.$$

le dernier terme, qui eſt 64, peut eſtre diuiſé ſans fraction par 1, 2, 4, 8, 16, 32, & 64; C'eſt pourquoy il faut examiner par ordre ſi cete Equation ne peut point eſtre diuiſée par quelqu'vn des binomes, $yy - 1$ ou $yy + 1, yy - 2$ ou $yy + 2, yy - 4$ &c. & on trouue qu'elle peut l'eſtre par $yy - 16$, en cete ſorte.

$$
\begin{array}{l}
+ \ y^6 - 8y^4 - 124yy - 64 \ \infty \ 0 \\
- 1 \ y^6 - 8y^4 - \ 4yy \ \ - - \quad -16 \\
\hline
0 \ - \ 16y^4 - 128yy \\
\quad\quad\quad 16 \quad\quad 16 \\
\hline
+ \ y^4 + 8yy \ \ + 4 \ \ \infty 0.
\end{array}
$$

Ie commence par le dernier terme, & diuiſe -64 par -16, ce qui fait $+4$, que i'eſcris dans le quotient, puis ie multiplie $+4$ par $+yy$, ce qui fait $+4yy$; c'eſt pourquoy i'eſcris $-4yy$ en la ſomme, qu'il faut diuiſer. car il y faut

La façon de diuiſer vne Equation par vn binome qui contiét ſa racine.

Bbb 3

faut toufiours efcrire le figne $+$ ou $--$ tout contraire a
celuy que produiſt la multiplication.& ioignant $-- 124 yy$
auec $-- 4 yy$, iay $-- 128 yy$, que ie diuiſe derechef par $-- 16$,
& iay $+ 8 yy$, pour mettre dans le quotient & en le multipliant par yy,iay $-- 8 y^4$,pour ioindre auec le terme qu'il
faut diuiſer, qui eſt auſſy $-- 8 y^4$, & ces deux enſemble
font $-- 16 y^4$, que ie diuiſe par $-- 16$, ce qui fait $+ 1 y^4$
pour ſe quotient, & $-- 1 y_6$ pour ioindre auec $+ 1 y^6$, cequi fait o, & monſtre que la diuiſion eſt acheuée. Mais
s'il eſtoit reſté quelque quantité, oubien qu'on n'euſt pû
diuiſer ſans fraction quelqu'vn des termes precedens, on
euſt par la reconnu, quelle ne pouuoit eſtre faite.

T out de meſme ſi on a $y^6 {+aa \atop --2cc} y {--a^4 \atop +c^4} yy {--a^6 \atop -2a^4cc \atop --aac^4} \infty \; 0.$

le dernier terme ſe peut diuiſer ſans fraction par
a, aa, $aa + cc$, $a^3 + acc$, & ſemblables. Mais il n'y en a
que deux qu'on ait beſoin de conſiderer, a ſçauoir aa &
$aa + cc$; car les autres donnant plus ou moins de dimenſions dans le quotient, qu'il n'y en a en la quantité connuë du penultieſme terme, empeſcheroient que la diuiſion ne s'y pûſt faire. Et notés, que ie ne conte icy les
dimenſions d'y^6, que pour trois, a cauſe qu'il ny a point
d'y^5, ny d'y^3, ny d'y en toute la ſomme. Or en examinant le binôme $yy -- aa -- cc \infty \, 0$, on trouue que la diuiſion
ſe peut faire par luy en cete ſorte.

$$+ y^6 {+aa \atop --2cc} y^4 {--a^4 \atop +c^4} yy {--a^6 \atop --2a^4cc \atop --aac^4} \quad \infty \, 0,$$

$$\frac{-- y^6 {--2aa \atop +cc} \quad {--a^4 \atop --aacc \atop --aacc} \quad {} \atop --aa--cc}{+ y^4 {+2aa \atop --cc} yy {+a^4 \atop +aacc} \quad \infty \, 0.}$$

Ce-

write in the dividend (for the opposite sign from that obtained by the multiplication must always be used). Adding $-124y^2$ and $-4y^2$ I have $-128y^2$. Dividing this by -16 I have $+8y^2$ in the quotient, and multiplying by y^2 I have $-8y^4$ to be added to the corresponding term, $-8y^4$, in the dividend. This gives $-16y^4$ which divided by -16 yields $+y^4$ in the quotient and $-y^6$ to be added to $+y^6$ which gives zero, and shows that the division is finished.

If, however, there is a remainder, or if any modified term is not exactly divisible by 16, then it is clear that the binomial is not a divisor.[216]

Similarly, given

$$y^6 + \left.\begin{matrix} a^2 \\ - 2c^2 \end{matrix}\right\} y^4 - \left.\begin{matrix} a^4 \\ + c^4 \end{matrix}\right\} y^2 - \left.\begin{matrix} a^6 \\ - 2a^4c^2 \\ - a^2c^4 \end{matrix}\right\} = 0,$$

the last term is divisible by a, a^2, a^2+c^2, a^3+ac^2, and so on, but only two of these need be considered, namely a^2 and a^2+c^2. The others give a term in the quotient of lower or higher degree than the known quantity of the next to the last term, and thus render the division impossible.[217] Note that I am here considering y^6 as of the third degree, since there are no terms in y^5, y^3, or y. Trying the binomial

$$y^2 - a^2 - c^2 = 0$$

we find that the division can be performed as follows:

$$\begin{matrix} + y^6 + \left.\begin{matrix} a^2 \\ - y^6 - 2c^2 \end{matrix}\right\} y^4 - \left.\begin{matrix} a^4 \\ + c^4 \end{matrix}\right\} y^2 - \left.\begin{matrix} a^6 \\ - 2a^4c^2 \end{matrix}\right\} = 0 \\ \hline \begin{matrix} 0 - 2a^2 \\ + c^2 \end{matrix}\} y^4 - \left.\begin{matrix} a^4 \\ - a^2c^2 \end{matrix}\right\} y^2 - \left.\begin{matrix} a^2c^4 \\ - a^2 - c^2 \end{matrix}\right\} \\ \hline \begin{matrix} - a^2 - c^2 \end{matrix} \quad - a^2 - c^2 \\ \hline + y^4 \quad + \left.\begin{matrix} 2a^2 \\ - c^2 \end{matrix}\right\} y^2 + \left.\begin{matrix} a^4 \\ + a^2c^2 \end{matrix}\right\} = 0, \end{matrix}$$

[216] This is evidently a modified form of our modern "synthetic division," the basis of our "Remainder Theorem," and of Horner's Method of solving numerical equations, a method known to the Chinese in the thirteenth century. See Cantor, Vol. II(1), pp. 279 and 287. See also Smith and Mikami, *History of Japanese Mathematics*, Chicago, 1914; Smith, I, 273.

[217] This is not a general rule.

This shows that $a^2 + c^2$ is the required root, which can easily be proved by multiplication.

But when no binomial divisor of the proposed equation can be found, it is certain that the problem depending upon it is solid,[218] and it is then as great a mistake to try to construct it by using only circles and straight lines as it is to use the conic sections to construct a problem requiring only circles; for any evidence of ignorance may be termed a mistake.

Again, given an equation in which the unknown quantity has four dimensions.[219] After removing any surds or fractions, see if a binomial having one term a factor of the last term of the expression will divide the left member. If such a binomial can be found, either the known quantity of the binomial is the required root, or,[220] after the division is performed, the resulting equation, which is of only three dimensions, must be treated in the same way. If no such binomial can be found, we must increase or diminish the roots so as to remove the second term, in the way already explained, and then reduce it to another of the third degree, in the following manner: Instead of

$$x^4 \pm px^2 \pm qx \pm r = 0$$

write

$$y^6 \pm 2py^4 + (p^2 \pm 4r)y^2 - q^2 = 0.[221]$$

[218] That is, that it involves a conic or some higher curve.

[219] A biquadratic equation.

[220] "Either, or," as in the original. It is like saying that the root of $x^2 - a^2 = 0$ is either $x = a$ or $x = -a$.

[221] Descartes wrote substantially "Instead of

$$+ x^{4*}.pxx.qx.r \infty 0$$

write

$$+ y^6.2py^4 + (pp.4r)yy - qq \infty 0."$$

The symbolism is characteristic of Descartes.

Ce qui monftre que la racine cherchée eft $aa + cc$.
Et la preuue en eft ayfée a faire par la multiplication.

Mais lorfqu'on ne trouue aucun binóme, qui puiffe Quels problef-mes font folides, lorfque l'Equa-tion eft cubique ainfi diuifer toute la fomme de l'Equation propofee, il eft certain que le Problefme qui en depend eft folide. Et ce n'eft pas vne moindre faute aprés cela, de tafcher a le conftruire fans y employer que des cercles & des lignes droites, que ce feroit d employer des fections coniques a conftruire ceux aufquels on n'a befoin que de cercles. car enfin tout ce qui tefmoigne quelque ignorance s'ap-pele faute.

Que fi on a vne Equation dont la quantité inconnuë La redu-ction des Equa-tions qui ont qua-tre di-méfions, lorfque le problef-me eft plan. Et quels fonr ceux qui font foli-des. ait quatre dimenfions, il faut en mefme façon , aprés en auoir ofté les nombres fours, & rompus, s'il y en a, voir fi on pourra trouuer quelque binóme, qui diuife toute la fomme, en le compofant de l'vne des quantités , qui di-uifent fans fraction le dernier terme. Et fi on en trouue vn, oubien la quantité connuë de ce binóme eft la racine cherchée; on du moins aprés cete diuifion, il ne refte en l'Equation, que trois dimenfions , en fuite dequoy il faut derechef l'examiner en la mefme forte. Mais lorf-qu'il ne fe trouue point de tel binóme , il faut en au-gmentant, ou diminuant la valeur de la racine, ofter le fecond terme de la fomme , en la façon tantoft expli-quée. Et aprés la reduire a vne autre , qui ne contie-ne que trois dimenfions. Cequi fe fait en cete forte.

Au lieu de $+ x^4 \overset{*}{.} p x x . q x . r \ \infty \ 0,$

il faut efcrire $+ y^6 . 2 p y^4 \overset{+ p p}{\underset{- 4 r}{}} y y - q q \ \infty \ 0.$

Et pour les fignes $+$ ou $--$ que iay omis, s'il y a
eu

eu $+ p$ en la precedente Equation, il faut mettre en cellecy $+ 2 p$, ou s'il y a eu $- p$, il faut mettre $-- 2 p$. & au contraire s'il y a eu $+ r$, il faut mettre $-- 4 r$, ou s'il y a eu $-- r$, il faut mettre $+ 4 r$. & foit qu'il y ait eu $+ q$, ou $-- q$, il faut toufiours mettre $-- qq$, & $+ pp$. au moins fi on fuppofe que x^4, & y^6 font marqués du fignes $+$, car ce feroit tout le contraire fi on y fuppofoit le figne $--$.

Par exemple fi on a $+ x^4 * -- 4 x x -- 8 x + 35 \infty 0$ il faut efcrire en fon lieu $y^6 -- 8 y^4 -- 124 yy -- 64 \infty 0$. car la quantité que iay, nommée p eftant $-- 4$, il faut mettre $-- 8 y^4$ pour $2 p y^4$. & celle, que iay nommée r eftant 35, il faut mettre $\overset{+\,16}{\underset{--\,140}{}} yy$, c'eft a dire $-- 124 yy$, au lieu de $\overset{+\,pp}{\underset{--\,4r}{}} yy$. & enfin q eftant 8, il faut mettre $-- 64$, pour $-- qq$. Tout de mefme au lieu de $+ x^4 * -- 17 xx -- 20 x -- 6 \infty 0$. il faut efcrire　　$+ y^6 -- 34 y^4 + 313 yy -- 400 \infty 0$. Car 34 eft double de 17, & 313 en eft le quarré ioint au quadruple de 6, & 400 eft le quarré de 20.

Tout de mefme auffy au lieu de

$$+ z^4 \cdot \overset{+\frac{1}{2}aa}{\underset{--\ cc}{}} zz \overset{--a^3}{\underset{--acc}{}} z \overset{+\frac{5}{16}a^4}{\underset{--\frac{1}{4}aacc}{}} \infty 0,$$

Il faut efcrire

$$y^6 \overset{+aa}{\underset{--2cc}{}} y^4 \overset{--a^4}{\underset{+c^4}{}} yy \overset{--a^6}{\underset{--2a^4cc}{\underset{--aac^4}{}}} \infty 0.$$

Car p eft $+ \frac{1}{2} aa -- cc$, & pp, eft $\frac{1}{4} a^4 -- aacc + c^4$, & $4 r$ eft $-- \frac{5}{4} a^4 + aacc$, & enfin $-- qq$ eft $-- a^6 -- 2 a^4 cc -- aac^4$.

Aprés que l'Equation eft ainfi reduite a trois dimenfions, il faut chercher la valeur d'yy par la methode defia expliquée; Et fi celle ne peut eftre trouuée, on n'a point befoin

For the ambiguous[222] sign put $+2p$ in the second expression if $+p$ occurs in the first; but if $-p$ occurs in the first, write $-2p$ in the second; and on the contrary, put $-4r$ if $+r$, and $+4r$ if $-r$ occurs; but whether the first expression contains $+q$ or $-q$ we always write $-q^2$ and $+p^2$ in the second, provided that x^4 and y^6 have the sign $+$; otherwise, we write $+q^2$ and $-p^2$. For example, given

$$x^4 - 4x^2 - 8x + 35 = 0$$

replace it by

$$y^6 - 8y^4 - 124y^2 - 64 = 0.$$

For since $p = -4$, we replace $2py^4$ by $-8y^4$; and since $r = 35$, we replace $(p^2-4r)y^2$ by $(16-140)y^2$ or $-124y^2$; and since $q = 8$, we replace $-q^2$ by -64.

Similarly, instead of

$$x^4 - 17x^2 - 20x - 6 = 0$$

we must write

$$y^6 - 34y^4 + 313y^2 - 400 = 0,$$

for 34 is twice 17, and 313 is the square of 17 increased by four times 6, and 400 is the square of 20.

In the same way, instead of

$$+ z^4 + \left(\frac{1}{2} a^2 - c^2\right) z^2 - (a^3 + ac^2)z - \frac{5}{16} a^4 - \frac{1}{4} a^2c^2 = 0,$$

we must write

$$y^6 + (a^2 - 2c^2)y^4 + (c^4 - a^4)y^2 - a^6 - 2a^4c^2 - a^2c^4 = 0;$$

for

$$p = \frac{1}{2} a^2 - c^2, \; p^2 = \frac{1}{4} a^4 - a^2c^2 + c^4, \; 4r = -\frac{5}{4} a^4 + a^2c^2.$$

And, finally,

$$- q^2 = - a^6 - 2a^4c^2 - a^2c^4.$$

When the equation has been reduced to three dimensions, the value of y^2 is found by the method already explained. If this cannot be

[222] Descartes wrote "pour les signes $+$ ou $-$ que j'ai omis."

done it is useless to pursue the question further, for it follows inevitably that the problem is solid. If, however, the value of y^2 can be found, we can by means of it separate the preceding equation into two others, each of the second degree, whose roots will be the same as those of the original equation. Instead of $+ x^4 \pm px^2 \pm qx \pm r = 0$, write the two equations

$$+ x^2 - yx + \frac{1}{2} y^2 \pm \frac{1}{2} p \pm \frac{q}{2y} = 0$$

and

$$+ x^2 + yx + \frac{1}{2} y^2 \pm \frac{1}{2} p \pm \frac{q}{2y} = 0.$$

For the ambiguous signs write $+ \frac{1}{2} p$ in each new equation, when p has a positive sign, and $- \frac{1}{2} p$ when p has a negative sign, but write $+ \frac{q}{2y}$ when we have $-yx$, and $- \frac{q}{2y}$ when we have $+yx$, provided q has a positive sign, and the opposite when q has a negative sign. It is then easy to determine all the roots of the proposed equation, and consequently to construct the problem of which it contains the solution, by the exclusive use of circles and straight lines. For example, writing $y^6 - 34y^4 + 313y^2 - 400 = 0$ instead of $x^4 - 17x^2 - 20x - 6 = 0$ we find that $y^2 = 16$; then, instead of the original equation

$$+ x^4 - 17x^2 - 20x - 6 = 0$$

write the two equations $+ x^2 - 4x - 3 = 0$ and $+ x^2 + 4x + 2 = 0$.

For, $y = 4$, $\frac{1}{2} y^2 = 8$, $p = 17$, $q = 20$, and therefore

$$+ \frac{1}{2} y^2 - \frac{1}{2} p - \frac{q}{2y} = -3$$

and

$$+ \frac{1}{2} y^2 - \frac{1}{2} p + \frac{q}{2y} = +2.$$

befoin de paffer outre; car il fuit de là infalliblement,
que le problefme eft folide. Mais fi on la trouue , on
peut diuifer par fon moyen la precedente Equation en
deux antres , en chafcune defquelles la quantité inconnuë n aura que deux dimenfions , & dont les racines feront les mefmes que les fienes. A fçauoir, au lieu de

$$+ x \cdot {}^{*} . p\,x\,x . q\,x . \ r \infty\, 0,$$

il faut efcrire ces deux autres

$$+ x\,x - y\,x + \tfrac{1}{2}yy . \tfrac{1}{2}p . \frac{q}{2y} \ \infty\, 0, \&$$

$$+ xx + y\,x + \tfrac{1}{2}yy . \tfrac{1}{2}p . \frac{q}{2y} \ \infty\, 0.$$

Et pour les fignes + & -- que iay omis, s'il y a + p en
l'Equation precedente, il faut mettre + $\tfrac{1}{2}p$ en chafcune
de celles cy; & -- $\tfrac{1}{2}p$, s'il y a en l'autre -- p. Mais il faut
mettre + $\frac{q}{2y}$, en celle où il y a -- $y\,x$; & -- $\frac{q}{2y}$, en celle où il
y a + $y\,x$, lorfqu'il y a + q en la premiere. Et au con-
traire s'il y a -- q, il faut mettre -- $\frac{q}{2y}$, en celle. où il y a
-- $y\,x$; & + $\frac{q}{2y}$, en celle où il y a + $y\,x$. En fuite dequoy
il eft ayfé de connoiftre toutes les racines de l'Equation
propofée, & par confequent de conftruire le problefme,
dont elle contient la folution, fans y employer que des
cercles, & des lignes droites.

Par exemple a caufe que faifant

$$y^{6} -- 34 y^{4} + 313\,yy - 400 \infty\, 0, \text{ pour}$$

$$x \cdot {}^{*} -- 17\,xx -- 20\,x -- 6 \infty\, 0, \text{ on trouue que } yy \text{ eft } 16, \text{ on-}$$

doii au lieu de cete Equation

$$+ x^{4}\,{}^{*} -- 17\,xx \cdot\!\cdot -- 20\,x -- 20\,x -- 6 \infty\, 0, \text{ efcrire ces deux}$$

C c c   autres

autres $+ xx - 4x - 3 \infty 0$. Et $+ xx + 4x + 2 \infty 0$.

car y eſt 4, $\frac{1}{2}yy$ eſt 8, p eſt 17, & q eſt 20, de façonque $+\frac{1}{2}yy -\frac{1}{2}p -\frac{q}{2y}$ fait $- 3$, & $+\frac{1}{2}yy -\frac{1}{2}p +\frac{q}{2y}$ fait $+ 2$. Et tirant les racines de ces deux Equations, on trouue toutes les meſmes, que ſi on les tiroit de celle où eſt x^4, a ſçauoir on en trouue vne vraye, qui eſt $\sqrt{7} + 2$, & trois fauſſes, qui ſont $\sqrt{7} - 2$, $2 + \sqrt{2}$, & $2 - \sqrt{2}$.

Ainſi ayant $x^4 - 4xx - 8x + 35 \infty 0$, pourceque la racine de $y^6 - 8y^4 - 124yy + 64 \infty 0$, eſt derechef 16, il faut eſcrire

$xx - 4x + 5 \infty 0$, & $xx + 4x + 7 \infty 0$.

Car icy $+\frac{1}{2}yy -\frac{1}{2}p -\frac{q}{2y}$ fait 5, & $+\frac{1}{2}yy -\frac{1}{2}p +\frac{q}{2y}$ fait 7. Et pourcequ'on ne trouue aucune racine, ny vraye, ny fauſſe, en ces deux dernieres Equations, on connoiſt de là que les quatre de l'Equation dont elles procedent ſont imaginaires; & que le Probleſme, pour lequel on l'a trouuée, eſt plan de ſa nature; mais qu'il ne ſçauroit en aucune façon eſtre conſtruit, a cauſe que les quantités données ne peuuent ſe ioindre.

Tout de meſme ayant

$$z^{4*} \begin{matrix} +\frac{1}{2}aa \\ -cc \end{matrix} \bigg\} zz \begin{matrix} -a^3 \\ -acc \end{matrix} \bigg\} z \begin{matrix} +\frac{5}{16}a^4 \\ -\frac{1}{4}aacc \end{matrix} \infty 0,$$

pourcequ'on trouue $aa + cc$ pour yy, il faut eſcrire

$zz - \sqrt{aa + cc}\, z + \frac{2}{4}aa - \frac{1}{2}a\sqrt{aa + cc} \infty 0$, &
$zz + \sqrt{aa + cc}\, z + \frac{2}{4}aa + \frac{1}{2}a\sqrt{aa + cc} \infty 0$.

Car y eſt $\sqrt{aa + cc}$, & $+\frac{1}{2}yy + \frac{1}{2}p$ eſt $\frac{2}{4}aa$, & $\frac{q}{2y}$ eſt $\frac{1}{2}a\sqrt{aa + cc}$. D'où on connoiſt que la valeur de z eſt.

Obtaining the roots of these two equations, we get the same results as if we had obtained the roots of the equation containing x^4, namely, one true root, $\sqrt{7} + 2$, and three false ones, $\sqrt{7} - 2$, $2 + \sqrt{2}$, and $2 - \sqrt{2}$. Again, given $x^4 - 4x^2 - 8x + 35 = 0$, we have $y^6 - 8y^4 - 124y^2 - 64 = 0$, and since the root of the latter equation is 16, we must write $x^2 - 4x + 5 = 0$ and $x^2 + 4x + 7 = 0$. For in this case,

$$+ \frac{1}{2} y^2 - \frac{1}{2} p - \frac{q}{2y} = 5$$

and

$$+ \frac{1}{2} y^2 - \frac{1}{2} p + \frac{q}{2y} = 7.$$

Now these two equations have no roots either true or false,[223] whence we know that the four roots of the original equation are imaginary; and that the problem whose solution depends upon this equation is plane, but that its construction is impossible, because the given quantities cannot be united.[224]

Similarly, given

$$z^4 + \left(\frac{1}{2} a^2 - c^2\right) z^2 - (a^3 + ac^2) z + \frac{5}{16} a^4 - \frac{1}{4} a^2 c^2 = 0,$$

since we have found $y^2 = a^2 + c^2$, we must write

$$z^2 - \sqrt{a^2 + c^2} z + \frac{3}{4} a^2 - \frac{1}{2} a \sqrt{a^2 + c^2} = 0,$$

and

$$z^2 + \sqrt{a^2 + c^2} z + \frac{3}{4} a^2 + \frac{1}{2} a \sqrt{a^2 + c^2} = 0.$$

[223] That is, all its roots are imaginary.
[224] That is, the given quantities cannot be taken together in the same problem.

For $y = \sqrt{a^2 + c^2}$ and $+\frac{1}{2}y^2 + \frac{1}{2}p = \frac{3}{4}a^2$, and $\frac{q}{2y} = \frac{1}{2}a\sqrt{a^2+c^2}$, then we have

$$z = \frac{1}{2}\sqrt{a^2 + c^2} + \sqrt{-\frac{1}{2}a^2 + \frac{1}{4}c^2 + \frac{1}{2}a\sqrt{a^2 + c^2}}$$

or

$$z = \frac{1}{2}\sqrt{a^2 + c^2} - \sqrt{-\frac{1}{2}a^2 + \frac{1}{4}c^2 + \frac{1}{2}a\sqrt{a^2 + c^2}}.$$

Now we already have $z + \frac{1}{2}a = x$, and therefore x, the quantity in the search for which we have performed all these operations, is

$$+\frac{1}{2}a + \sqrt{\frac{1}{4}a^2 + \frac{1}{4}c^2} - \sqrt{\frac{1}{4}c^2 - \frac{1}{2}a^2 + \frac{1}{2}a\sqrt{a^2 + c^2}}.$$

To emphasize the value of this rule, I shall apply it to a problem. Given the square AD and the line BN, to prolong the side AC to E, so that EF, laid off from E on EB, shall be equal to NB.

Pappus showed that if BD is produced to G, so that DG = DN, and a circle is described on BG as diameter, the required point E will be the intersection of the straight line AC (produced) with the circumference of this circle.[225]

Those not familiar with this construction would not be likely to discover it, and if they applied the method suggested here they would never think of taking DG for the unknown quantity rather than CF or FD, since either of these would much more easily lead to an equa-

---

[225] Pappus Lib. VII, Prop. 72, Vol. II, p. 783. The following is in substance the proof given by Pappus. He first gives an elaborate proof of the following lemma: Given a square ABCD, and E a point in AC produced, EG perpendicular to BE at E, meeting BD produced in G, and F the point of intersection of BE and CD. Then $\overline{CD}^2 + FE^2 = \overline{DG}.^2$ Then he proceeds as follows: By the construction given in the problem, $\overline{DN}^2 = BD^2 + \overline{BN}^2$. By the lemma, $\overline{DG}^2 = \overline{CD}^2 + \overline{FE}^2$. By construction, BD = CD and DG = DN. Therefore, FE = BN.

eſt $\frac{1}{2}.\sqrt{aa + cc} + \sqrt{--\frac{1}{2}aa + \frac{1}{4}cc} + \frac{1}{2}a\sqrt{aa + cc}$,

oubien $\frac{1}{2}\sqrt{aa + cc} - \sqrt{--\frac{1}{2}aa + \frac{1}{4}cc} + \frac{1}{2}a\sqrt{aa + cc}$.

Et pourceque nous auions fait cy deſſus $\approx + \frac{1}{2}a \infty x$, nous apprenons que la quantité x, pour la connoiſſance de laquelle nous auons fait toutes ces operations, eſt

$$-+ \tfrac{1}{2}a + \sqrt{\tfrac{1}{4}aa + \tfrac{1}{4}cc} - \sqrt{\tfrac{1}{4}cc - \tfrac{1}{2}aa + \tfrac{1}{2}a\sqrt{aa + cc}}.$$

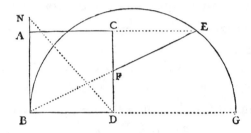

Exemple de l'vſage de ces reductions.

Mais affin qu'on puiſſe mieux connoiſtre l'vtilité de cete reigle il faut que ie l'applique a quelq; Probleſme. Si le quarré A D, & la ligne B N eſtant donnés, il faut prolonger le coſté A C iuſques a E, en ſorte qu'E F, tirée d'E vers B, ſoit eſgale a N B. On apprent de Pappus, qu'ayant premierement prolongé B D iuſques à G , en ſorte que D G ſoit eſgale à D N, & ayant deſcrit vn cercle dont le diametre ſoit B G , ſi on prolonge la ligne droite A C, elle rencontrera la circonference de ce cercle au point E, qu'on demandoit. Mais pour ceux qui ne ſçauroiet point cete côſtruction elle ſeroit aſſés difficile à rencôtrer, & en la cherchât par la methode icy propoſée, ils ne s'auiſeroiët iamais de prêdre D G pour la quãtité inconnuë, mais plutoſt C F, ou F D, a cauſe que ce

Ccc 2 ſont

font elles qui conduifent le plus ayfement a l'Equatiõ: &
lors ils en trouueroiẽt vne qui ne feroit pas facile a deme-
fler, fans la reigle que ie viens d'expliquer. Car pofant a
pour B D ou C D, & c pour E F, & x pour D F, on a C F
$\infty a - x$, & cõme C F ou $a - x$, eft à F E ou c, ainfi F D ou x,
eft a B F, qui par confequent eft $\frac{cx}{a - x}$. Puis a caufe du triangle rectangle B D F, dont les coftés font l'vn x & l'autre a, leurs quarrés, qui font $xx + aa$, font efgaux a celuy de la baze, qui eft $\frac{ccxx}{xx - 2ax + aa}$, de façon que multipliant le tout par $xx - 2ax + aa$, on trouue que l'E-
quation eft $x^4 - 2ax^3 + 2aaxx - 2a^3x + a^4 \infty ccxx$,
oubien $x^4 - 2ax^3 {+2aa \atop -cc} xx - 2a^3x + a^4 \infty o$. Et on
connoift par les reigles precedentes, que fa racine, qui
eft la longeur de la ligne D F, eft $\frac{1}{2}a + \sqrt{\frac{1}{4}aa + \frac{1}{4}cc}$
$- \sqrt{\frac{1}{4}cc - \frac{1}{2}aa + \frac{1}{2}a\sqrt{aa + cc}}$.

Que fi on pofoit B F, ou C E, ou B E pour la quantité
inconnuë, on viendroit derechef à vne Equation, en laquelle il y auroit 4 dimenfions, mais qui feroit plus ayfée
a démefler, & on y viendroit affés ayfement; au lieu que
fi c'eftoit D G qu'on fuppofaft, on viendroit beaucoup
plus difficilement a l'Equation, mais auffy elle feroit tres
fimple. Ce que ie mets icy pour vous auertir, que lorfque le Problefme propofé n'eft point folide, fi en le cherchant par vn chemin on vient a vne Equation fort com-
pofee, on peut ordinairement venir a vne plus fimple, en
le cherchant par vn autre.

Ie pourrois encore aioufter diuerfes reigles pour démefler les Equations, qui vont au Cube, ou au Quarre
de

tion. They would thus get an equation which could not easily be solved without the rule which I have just explained.

For, putting a for BD or CD, c for EF and x for DF, we have CF $= a - x$, and, since CF is to FE as FD is to BF, we have

$$a - x : c = x : \text{BF},$$

whence BF $= \dfrac{cx}{a-x}$. Now, in the right triangle BDF whose sides are x and a, $x^2 + a^2$, the sum of their squares, is equal to the square of the hypotenuse, which is $\dfrac{c^2 x^2}{x^2 - 2ax + a^2}$ Multiplying both sides by

$$x^2 - 2ax + a^2$$

we get the equation,

$$x^4 - 2ax^3 + 2a^2 x^2 - 2a^3 x + a^4 = c^2 x^2,$$

or

$$x^4 - 2ax^3 + (2a^2 - c^2) x^2 - 2a^3 x + a^4 = 0,$$

and by the preceding rule we know that its root, which is the length of the line DF, is

$$\frac{1}{2} a + \sqrt{\frac{1}{4} a^2 + \frac{1}{4} c^2} - \sqrt{\frac{1}{4} c^2 - \frac{1}{2} a^2 + \frac{1}{2} a \sqrt{a^2 + c^2}}.$$

If, on the other hand, we consider BF, CE, or BE as the unknown quantity, we obtain an equation of the fourth degree, but much easier to solve, and quite simply obtained.[226]

Again, if DG were used, the equation would be much more difficult to obtain, but its solution would be very simple. I state this simply to warn you that, when the proposed problem is not solid, if one method of attack yields a very complicated equation a much simpler one can usually be found by some other method.

[226] Taking BF as the unknown quantity, the resulting equation is

$$x^4 + 2cx^3 + (c^2 - 2a^2) x^2 - 2a^2 cx - a^2 c^2 = 0.$$

Rabuel, p. 487.

I might add several different rules for the solution of cubic and biquadratic equations but they would be superfluous, since the construction of any plane problem can be found by means of those already given.

I could also add rules for equations of the fifth, sixth, and higher degrees, but I prefer to consider them all together and to state the following general rule:

First, try to put the given equation into the form of an equation of the same degree obtained by multiplying together two others, each of a lower degree. If, after all possible ways of doing this have been tried, none has been sucessful, then it is certain that the given equation cannot be reduced to a simpler one; and, consequently, if it is of the third or fourth degree, the problem depending upon it is solid; if of the fifth or sixth, the problem is one degree more complex, and so on. I have also omitted here the demonstration of most of my statements, because they seem to me so easy that if you take the trouble to examine them systematically the demonstrations will present themselves to you and it will be of much more value to you to learn them in that way than by reading them.

de quarré, mais elles feroient fuperfluës; car lorfque les Problefmes font plans, on en peut toufiours trouuer la conftruction par celles cy.

Ie pourrois auffy en adioufter d autres pour les Equations qui montent iufques au furfolide, ou au Quarré de cube, ou au delà, mais i'ayme mieux les comprendre toutes en vne, & dire en general, que lorfqu'on a tafché de les reduire a mefme forme, que celles d'autant de dimenfions, qui vienent de la multiplication de deux autres qui en ont moins, & qu'ayant dénombré tous les moyens, par lefquels cete multiplication eft poffible, la chofe n'a pû fucceder par aucun, on doit s'affurer qu'elles ne fçauroient eftre reduites a de plus fimples. En forte que fi la quantité inconnuë a 3 on 4 dimenfions, le Problefme pour lequel on la cherche eft folide; & fi elle en a 5, on 6, il eft d'vn degré plus compofé; & ainfi des autres.

Regle generale pour reduire les Equatiõs qui paffent le quarré de quarré.

Au refte i'ay omis icy les demonftrations de la plus part de ce que iay dit a caufe qu'elles m'ont femblé fi faciles, que pourvû que vous preniés la peine d'examiner methodiquement fi iay failly, elles fe prefenteront a vous d'elles mefme: & il fera plus vtile de les apprendre en cete façon, qu'en les lifant.

Or quand on eft affuré, que le Problefme propofé eft folide, foit que l'Equation par laquelle on le cherche monte au quarré de quarré, foit qu'elle ne monte que iufques au cube, on peut toufiours en trouuer la racine par l'vne des trois fections coniques, laquelle que ce foit ou mefme par quelque partie de l'vne d'elles, tant petite qu'elle puiffe eftre; en ne fe feruãt au refte que de lignes droites, & de cercles. Mais ie me contenteray icy de

Facon generale pour conftruire tous les problefmes folides. reduits a vne Equatiõ de trois ou quatre dimenfions.

Ccc 3   donner

donner vne reigle generale pour les trouuer toutes par le moyen d'vne Parabole, a caufe qu'elle eft en quelque façon la plus fimple.

Premierement il faut ofter le fecond terme de l'Equation propofée, s'il n'eft defia nul, & ainfi la reduire à telle forme, $z^3 \infty *.\, a\, p\, z.\, a\, aq$, fi la quantité inconnuë n'a que trois dimenfions, oubien à telle, $z^4 \infty *.\, a p z z.\, a a q z.$ $a^3 r$, fi elle en a quatre; oubien en prenant a pour l'vnité, à telle, $z^3 \infty *.\, p\, z.\, q$, & à telle

$z^4 \infty *.\, p z z.\, q z.\, r.$

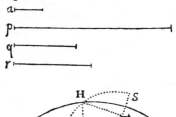

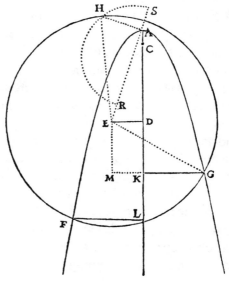

Aprés

Now, when it is clear that the proposed problem is solid, whether the equation upon which its solution depends is of the fourth degree or only of the third. its roots can always be found by any one of the three conic sections, or even by some part of one of them, however small, together with only circles and straight lines. I shall content myself with giving here a general rule for finding them all by means of a parabola, since that is in some respects the simplest of these curves.

First, remove the second term of the proposed equation. if this is not already zero, thus reducing it to the form $z^3 = \pm apz \pm a^2q$, if the given equation is of the third degree, or $z^4 = \pm apz^2 \pm a^2qz \pm a^3r$, if it is of the fourth degree. By choosing a as the unit, the former may be written

$z^3 = \pm pz \pm q$ and the latter $z^4 = \pm pz^2 \pm qz \pm r$. Suppose that the parabola FAG (pages 194-198) is already described; let ACDKL be its axis, a, or 1 which equals 2AC, its latus rectum (C being within the parabola), and A its vertex. Lay off CD equal to $\frac{1}{2}p$ so that the points D and A lie on the same side of C if the equation contains $+p$ and on opposite sides if it contains $-p$. Then at the point D (or, if $p = 0$, at C), erect DE perpendicular to CD, so that DE is equal to $\frac{1}{2}q$, and about E as center with AE as radius describe the circle FG, if the given equation is a cubic, that is, if r is zero.

Aprés cela fuppofant que la Parabole F A G eft defia
defcrite, & que fon aiffieu eft A C D K L, & que fon cofté droit eft *a*, ou 1, dont A C eft la moitié, & enfin que
le point C eft au dedans de cete Parabole, & que A en eft
le fommet; Il faut faire C D ∞ ½*p*, & la prendre du mefme cofté, qu'eft le point A au regard du point C, s'il y a
+ *p* en l'Equation; mais s'il y a -- *p* il faut la prendre de
l'autre cofte. Et du point D, oubien, fi la quantité

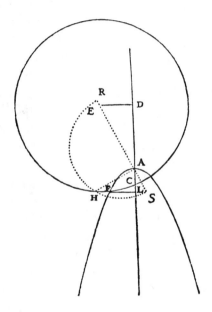

p eftoit nulle, du point C il faut efleuer vne ligne a angles droits iufques a E, en forte qu'elle foit efgale a ½ *q*.
Et enfin du centre E il faut defcrire le cercle F G, dont
le

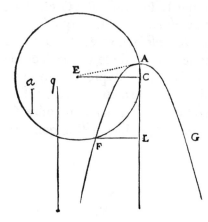

le demidiametre ſoit
A E , ſi l'Equation
n'eſt que cubique, en
ſorte que la quanti-
té *r* ſoit nulle. Mais
quand il y a ╶╂╴ *r* il
faut dans cete ligne
A E prolongée, pren-
dre d' vn coſté A R
eſgale à *r*, & de l'autre
A S eſgale au coſté
droit de la Parabole
qui eſt ɪ, & ayant de-
ſcrit vn cercle dont le diametre ſoit R S, il faut faire A H

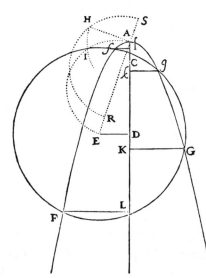

perpēdiculaire ſur
A E , laquelle A H
rencontre ce cer-
cle R H S au point
H, qui eſt celuy par
où l'autre cercle
F H G doit paſſer.
Et quand il y a ╶ *r*
il faut aprés auoir
ainſi trouué la ligne
A H, inſcrire A I,
qui luy ſoit eſgale,
dans vn autre cer-
cle, dont A E ſoit
le diametre, & lors
c'eſt par le point I,
que

If the equation contains $+ r$, on one side of AE produced, lay off AR equal to r, and on the other side lay off AS equal to the latus rectum of the parabola, that is, to 1, and describe a circle on RS as diameter. Then if AH is drawn perpendicular to AE it will meet the circle RHS in the point H, through which the other circle FHG must pass.

If the equation contains $- r$, construct a circle upon AE as diameter and in it inscribe AI, a line equal to AH;[227] then the first circle must pass through the point I.

[227] That is, draw a chord equal to AH.

Now the circle FG can cut or touch the parabola in 1, 2, 3, or 4 points; and if perpendiculars are drawn from these points upon the axis they will represent all the roots of the equation, both true and false. If the quantity q is positive the true roots will be those perpendiculars, such as FL, on the same side of the parabola, as E,[228] the center of the circle; while the others, as GK, will be the false roots. On the other hand, if q is negative, the true roots will be those on the opposite side, and the false or negative roots[229] will be those on the same side as E, the center of the circle. If the circle neither cuts nor touches the parabola at any point, it is an indication that the equation has neither a true nor a false root, but that all the roots are imaginary.[230]

This rule is evidently as general and complete as could possibly be desired. Its demonstration is also very easy. If the line GK thus constructed be represented by z, then AK is z^2, since by the nature of the parabola, GK is the mean proportional between AK and the latus rectum, which is 1. Then if AC or $\frac{1}{2}$, and CD or $\frac{1}{2}p$, be subtracted from AK, the remainder is DK or EM, which is equal to $z^2-\frac{1}{2}p-\frac{1}{2}$ of which the square is

$$z^4-pz^2-z^2+\frac{1}{4}p^2+\frac{1}{2}p+\frac{1}{4}.$$

And since $DE=KM=\frac{1}{2}q$, the whole line $GM=z+\frac{1}{2}q$, and the square

of GM equals $z^2+qz+\frac{1}{4}q^2$. Adding these two squares we have

$$z^4-pz^2+qz+\frac{1}{4}q^2+\frac{1}{4}p^2+\frac{1}{2}p+\frac{1}{4}$$

[228] That is, on the same side of the axis of the parabola.

[229] "Les fausses ou moindres que rien." This is the first time Descartes has directly used this synonym.

[230] It may be noted that Descartes considers the cubic as a quartic having zero as one of its roots. Therefore, the circle always cuts the parabola at the vertex. It must then cut it in another point, since the cubic must have one real root. It may or may not cut it in two other points. It may cut it in two coincident points at the vertex, in which case the equation reduces to a quadratic.

que doit paſſer F I G le premier cercle cherché. Or ce
cercle F G peut coupper, ou toucher la Parabole en 1,
ou 2, ou 3, ou 4 poins, deſquels tirant des perpendiculaires ſur laiſſieu, on a toutes les racines de l'Equation tant
vrayes, que fauſſes. A ſçauoir ſi la quantité q eſt marquée
du ſigne +, les vrayes racines ſeront celles de ces perpendiculaires, qui ſe trouueront du meſme coſté de la
parabole, que E le centre du cercle, comme F L ; & les
autres, comme G K, ſeront fauſſes : Mais au contraire ſi
cete quantité q eſt marquée du ſigne -- les vrayes ſeront
celles de l'autre coſté ; & les fauſſes, ou moindres que
rien ſeront du coſté où eſt E le centre du cercle. Et enfin ſi ce cercle ne couppe, ny ne touche la Parabole en aucun point, cela teſmoigne qu'il n'y a aucune racine ny
vraye ny fauſſe en l'Equation, & qu'elles ſont toutes
imaginaires. En ſorte que cete reigle eſt la plus generale, & la plus accomplie qu'il ſoit poſſible de ſouhaiter.

Et la demonſtration en eſt fort ayſée. Car ſi la ligne
G K, trouuée par cete conſtruction, ſe nomme z, A K
ſera zz, a cauſe de la Parabole, en laquelle G K doit
eſtre moyene proportionelle, entre A K, & le coſté droit
qui eſt 1. puis ſi de A K i'oſte A C, qui eſt $\frac{1}{2}$, & C D qui
eſt $\frac{1}{2}p$, il reſte D K, ou E M, qui eſt $zz -- \frac{1}{2}p -- \frac{1}{2}$, dont le
quarré eſt

$$z^4 -- pzz -- zz + \frac{1}{4}pp + \frac{1}{2}p + \frac{v}{4}.$$ & a cauſe que D E, ou
K M eſt $\frac{1}{2}q$, la toute G M eſt $z + \frac{1}{2}q$, dont le quarré eſt
$zz + qz + \frac{1}{4}qq$, & aſſemblant ces deux quarrés, on a
$z^4 -- pz + qz + \frac{1}{4}qq + \frac{1}{4}pp + \frac{1}{2}p + \frac{1}{4}$,

D d d   pour

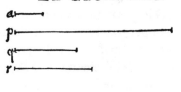

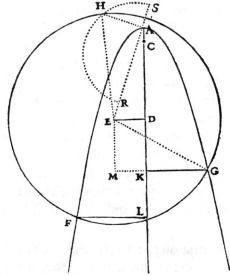

pour le quarré de la ligne G E, a caufe qu'elle eft la baze
du triangle rectangle E M G.

　　Mais a caufe que cete mefme ligne G E eft le demidiametre du cercle F G, elle fe peut encore expliquer en
d'autres termes, a fçauoir E D eftant $\frac{1}{2}q$, & A D eftant
$\frac{1}{2}p + \frac{1}{2}$, E A eft $\sqrt{\frac{1}{4}qq + \frac{1}{4}pp + \frac{1}{2}p + \frac{1}{4}}$ a caufe de l'angle droit A D E, puis H A eftant moyene proportionelle
entre A S qui eft 1 & A R qui eft r, elle eft $\sqrt{r}$. & à caufe de l'angle droit E A H, le quarré de H E, on E G eft
$\frac{1}{4}qq + \frac{1}{4}pp + \frac{1}{2}p + \frac{1}{4} + r$: fibienque il y a Equation
　　　　　　　　　　　　　　　　　　　　　　　　　　entre

for the square of GE, since GE is the hypotenuse of the right triangle EMG.

But GE is the radius of the circle FG and can therefore be expressed in another way. For since $ED = \frac{1}{2} q$, and $AD = \frac{1}{2} p + \frac{1}{2}$, and ADE is a right angle, we have

$$EA = \sqrt{\frac{1}{4} q^2 + \frac{1}{4} p^2 + \frac{1}{2} p + \frac{1}{4}}.$$

Then, since HA is the mean proportional between AS or 1 and AR or r, $HA = \sqrt{r}$; and since EAH is a right angle, the square of HE or of EG is

$$\frac{1}{4} q^2 + \frac{1}{4} p^2 + \frac{1}{2} p + \frac{1}{4} + r,$$

and we can form an equation from this expression and the one already obtained. This equation will be of the form $z^4 = pz^2 - qz + r$, and therefore the line GK, or z, is the root of this equation, which was to be proved. If you will apply this method in all the other cases, with the proper changes of sign, you will be convinced of its usefulness, without my writing anything further about it.

Let us apply it to the problem of finding two mean proportionals between the lines a and q. It is evident that if we represent one of the mean proportionals by z, then $a : z = z : \dfrac{z^2}{a} = \dfrac{z^2}{a} : \dfrac{z^3}{a^2}$. Thus we have an equation between q and $\dfrac{z^3}{a^2}$, namely, $z^3 = a^2 q$.

Describe the parabola FAG with its axis along AC, and with AC equal to $\frac{1}{2} a$, that is, to half the latus rectum. Then erect CE equal to $\frac{1}{2} q$ and perpendicular to AC at C, and describe the circle AF

entre cete fomme & la precedente. cequi eſt le meſme
que $z^4 \infty * p z z -- q z + r.$ & par conſequent la ligne tróu-
uee GK qui a eſté nommée z eſt la racine de cete Equation. ainſi qu'il falloit demonſtrer. Et ſi vous appliqués
ce meſme calcul a tous les autres cas de cete reigle, en
changeant les ſignes -+ & -- ſelon l'occaſion , vous y
trouuerés voſtre conte en meſme ſorte,ſans qu'il ſoit be-
tſoin que ie m'y areſte.

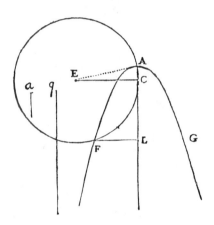

Si on veut donc ſuiuant cete reigle trouuer deux moyennes proportionelles entre les lignes a & q; chaſcun
ſçait que poſant z pour l'vne , comme a eſt à z , ainſi
$\overline{z}$ à $\frac{z z}{a}$, & $\frac{z z}{a}$ à $\frac{z^3}{a a}$; de façon qu'il y a Equation entre q &
$\frac{z^3}{a a}$, c'eſt a dire, $z^3 \infty * * a a q.$ Et la Parabole F A G eſtant

L'inuen-
tion de
deux mo-
yenes pro-
portio-
nelles.

Ddd 2   de-

defcrite, auec la partie de fon aiffieu A C, qui eft $\frac{1}{2}$ *a* la
moitié du cofté droit ; il faut du point C efleuer la perpendiculaire C E efgale à $\frac{1}{2}$ *q*, & du centre E, par A, de-
fcriuant le cercle A F, on trouue F L, & L A, pour les
deux moyennes cherchées.

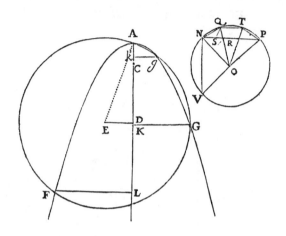

La facon
de diuifer
vn angle
en trois. Tout de mefme fi on veut diuifer l'angle N O P, oubien l'arc, ou portion de cercle N Q T P, en trois parties efgales; faifant N O ∞ 1, pour le rayon du cercle, &
N P ∞ *q*, pour la fubtendue de l'arc donné, & N Q ∞ *z*,
pour la fubtendue du tiers de cet arc ; l'Equation
vient,

$z^3 \infty * 3 z -- q$. Car ayant tiré les lignes N Q, O Q,
O T; & faifant Q S parallele a T O, on voit que comme
N O eft a N Q, ainfi N Q a Q R, & Q R a R S ; en forte
que

about E as center, passing through A. Then FL and LA are the required mean proportionals.[231]

Again, let it be required to divide the angle NOP, or rather, the circular arc NQTP, into three equal parts. Let $NO = 1$ be the radius of the circle, $NP = q$ be the chord subtending the given arc, and $NQ = z$ be the chord subtending one-third of that arc; then the equation is $z^3 = 3z - q$. For, drawing NQ, OQ and OT, and drawing QS parallel to TO, it is obvious that NO is to NQ as NQ is to QR as QR is to RS. Since $NO = 1$ and $NQ = z$, then $QR = z^2$ and $RS = z^3$; and since NP or q lacks only RS or z^3 of being three times NQ or z, we have $q = 3z - z^3$ or $z^3 = 3z - q$.[232]

Describe the parabola FAG so that CA, one-half its latus rectum, shall be equal to $\frac{1}{2}$; take $CD = \frac{3}{2}$ and the perpendicular $DE = \frac{1}{2} q$; then describe the circle FAgG about E as center, passing through A. This circle cuts the parabola in three points, F, g, and G, besides the vertex, A. This shows that the given equation has three roots, namely, the two true roots, GK and gk, and one false root, FL.[233] The smaller

[231] This may be shown as follows: Draw FM $\perp$ to EC; let FL=z. From the nature of the parabola, $\overline{FL}^2 = a$. AL; $AL = \frac{z^2}{a}$; $\overline{EC}^2 + \overline{CA}^2 = \overline{EA}^2$; $\overline{EM}^2 + \overline{FM}^2$

$= \overline{EF}^2$; $\overline{EA}^2 = \frac{q^2}{4} + \frac{a^2}{4}$; $\overline{EM}^2 = (EC - FL)^2 = \left(\frac{1}{2} q - z\right)^2$; $\overline{FM}^2 = \overline{CL}^2 = (AL - AC)^2$

$= \left(\frac{z^2}{a} - \frac{a}{2}\right)^2$; $\overline{EF}^2 = \frac{q^2}{4} - qz + z^2 + \frac{z^4}{a^2} - z^2 + \frac{a^2}{4}$. But EF=EA.

$$\therefore \frac{q^2}{4} + \frac{a^2}{4} = \frac{q^2}{4} - qz + z^2 + \frac{z^4}{a^2} - z^2 + \frac{a^2}{4},$$

whence $z^3 = a^2 q$.

[232] $\angle$ NOQ is measured by arc NQ;
$\angle$ QNS is measured by $\frac{1}{2}$ arc QP or arc NQ;
$\angle$ SQR=$\angle$ QOT is measured by arc QT or NQ;
$\therefore \angle$ OQN = $\angle$ NQR = $\angle$ QSR.
$\therefore$ NO : NQ = NQ : QR = QR : RS.
QR = z^2; RS = z^3. Let OT cut NP at M.
NP = 2NR + MR = 2NQ + MR
    = 2NQ + MS — RS
    = 2NQ + QT — RS
    = 3NQ — RS.
    Or $q = 3z - z^3$.
Rabuel, p. 534.
[233] G and g being on the opposite side of the axis from E, and F being on the same side.

of the two roots, *gk*, must be taken as the length of the required line
NQ, for the other root, GK, is equal to NV, the chord subtended by
one-third the arc VNP,[234] which, together with the arc NQP consti-
tutes the circle; and the false root, FL, is equal to the sum of QN and
NV, as may easily be shown.[235]

It is unnecessary for me to give other examples here, for all problems that are only solid can be reduced to such forms as not to require
this rule for their construction except when they involve the finding
of two mean proportionals or the trisection of an angle. This will be
obvious if it is noted that the most difficult of these problems can be'

[234] For proof, see Rabuel, page 535.

[235] Let $AB = b$; $EB = MR = mk = NL = c$; $AK = t$; $Ak = s$; $AL = r$;
$KG = y$; $kg = z$, $FL = v$. Then $GM = y + c$, $gm = z + c$, $FN = v - c$, $GK^2 = a \cdot AK$,

$at = y^2$, $t = \dfrac{y^2}{a}$, $\overline{gk}^2 = a \cdot Ak$, $as = z^2$, $s = \dfrac{z^2}{a}$, $\overline{FL}^2 = a \cdot AL$, $ar = v^2$, $r = \dfrac{v^2}{a}$,

$$ME = AB - AK = b - \frac{y^2}{a}$$

$mE = b - \dfrac{z^2}{a}$

$$EN = \frac{v^2}{a} - b$$

$$\overline{EG}^2 = \overline{EM}^2 + \overline{MG}^2$$

$$\overline{EA}^2 = \overline{AB}^2 + BE^2$$

$$\overline{EG}^2 = b^2 - \frac{2by^2}{a} + \frac{y^4}{a^2} + y^2 + 2cy + c^2$$

$2ab = \dfrac{y^3 + 2a^2c + a^2y}{y}$

$2ab = \left| \dfrac{z^3 + 2a^2c + a^2z}{z} \right.$

$$\frac{y^3 + 2a^2c + a^2y}{y} = \frac{z^3 + 2a^2c + a^2z}{z}$$

$$2a^2c = z^2y + zy^2$$

Similarly,

$$2a^2c = v^2y - vy^2$$

$z^2y + zy^2 = v^2y - vy^2$

$v^2 - z^2 = vy + zy$

$v - z = y$

$v = y + z$

$FL = KG + kg$

Rabuel, p. 540.

208

que N O eſtant 1, & N Q eſtant χ, Q R eſt $\chi\chi$, & R S eſt
χ³: Et a cauſe qu'il s'en faut ſeulement R S, ou χ³, que la
ligne N P, qui eſt q, ne ſoit triple de N Q, qui eſt χ, ou
à $q \infty 3 \chi - \chi$³ oubien,

$$\chi^3 \infty {}^* 3\chi - q.$$

Puis la Parabole F A G eſtant deſcrite, & C A la moi-
tié de ſon coſté droit principal eſtant $\frac{1}{2}$, ſi on prent C D
$\infty \frac{3}{2}$, & la perpendiculaire D E $\infty \frac{1}{2} q$, & que du centre E,
par A, on deſcriue le cercle F A g G, il couppe cete Pa-
rabole aux trois poins F, g, & G, ſans conter le point A
qui en eſt le ſommet. Ce qui monſtre qu'il y a trois racines en cete Equation, à ſçauoir les deux G K, & g k, qui
ſont vrayes; & la troiſieſme qui eſt fauſſe, á ſçauoir F L.
Et de ces deux vrayes c'eſt g k la plus petite qu'il faut
prendre pour la ligne N Q qui eſtoit cherchée. Car l'autre G K, eſt eſgale à N V, la ſubtendue de la troiſieſme
partie de l'arc N V P, qui auec l'autre arc N Q P acheue
le cercle. Et la fauſſe F L eſt eſgale a ces deux enſemble
Q N & N V, ainſi qu'il eſt ayſé a voir par le calcul.

Il ſeroit ſuperflus que ie m'areſtaſſe a donner icy d'au- Que tous les problesmes ſolides ſe peuuent reduire a ces deux conſtruc- tions.
tres exemples; car tous les Problesmes qui ne ſont que
ſolides ſe peuuent reduire a tel point, qu'on n'a aucun beſoin de cete reigle pour les conſtruire, ſinon entant qu'elle ſert a trouuer deux moyennes proportionelles, oubien
a diuiſer vn angle en trois parties eſgales. Ainſi que vous
connoiſtrés en conſiderant, que leurs difficultés peuuent
touſiours eſtre compriſes en des Equations, qui ne montent que iuſque au quarré de quarré, ou au cube : Et que
toutes celles qui montent au quarré de quarré, ſe reduiſent au quarré, par le moyen de quelques autres, qui ne

montent que iufques au cube: Et enfin qu'oǹ peut oſter
le ſecond terme de celles cy. En ſorte qu'il n'y en a point
qui ne ſe puiſſe reduire a quelq; vne de ces trois formes.

$$z^3 \infty * -- p\,z + q.$$
$$z^3 \infty_{\text{,}} * + p\,z + q.$$
$$z^3 \infty * + p\,z -- q.$$

Or ſi on a $z^3 \infty * -- p\,z + q$, la reigle dont Cardan at-
tribue l'inuention a vn nommé Scipio Ferreus, nous apprent que la racine eſt,

$$\sqrt{C. + \tfrac{1}{2}q + \sqrt{\tfrac{1}{4}qq + \tfrac{1}{27}p^3}} \,-\!\!\times\, \sqrt{C. \!\!\times\! \tfrac{1}{2}q + \sqrt{\tfrac{1}{4}qq + \tfrac{1}{27}p^3}}$$

Comme auſſy lorſqu'on a $z^3 \infty * + p\,z + q$, & que le
quarré de la moitié du dernier terme eſt plus grand que
le cube du tiers de la quantité connuë du penultiefme,
vne pareille reigle nous apprent que la racine eſt,

$$C. + \tfrac{1}{2}q + \sqrt{\tfrac{1}{4}qq -- \tfrac{1}{27}p^3} \,+\, \sqrt{C. + \tfrac{1}{2}q -- \sqrt{\tfrac{1}{4}qq -- \tfrac{1}{27}p^3}}$$

D'où il paroiſt qu'on peut conſtruire tous les Problefmes, dont les difficultés ſe reduiſent a l'vne de ces deux
formes, ſans auoir beſoin des ſections coniques pour autre choſe, que pour tirer les racines cubiques de quelques quantités données, c'eſt a dire, pour trouuer deux
moyennes proportionelles entre ces quantités & l'vnité.

Puis ſi on a $z^3 \infty * + p\,z + q$, & que le quarré de la
moitié du dernier terme ne ſoit point plus grand que le
cube du tiers de la quantité connuë du penultieſme, en
ſuppoſant le cercle N Q P V, dont le demidiametre N O
ſoit $\sqrt{\tfrac{1}{3}p}$, c'eſt a dire la moyenne proportionelle entre
le tiers de la quantité donnée p & l'vnité; & ſuppoſant
auſſy la ligne N P iuſcrite dans ce cercle qui ſoit $\frac{3q.}{p}$

c'eſt

expressed by equations of the third or fourth degree; that all equations of the fourth degree can be reduced to quadratic equations by means of other equations not exceeding the third degree; and finally, that the second terms of these equations can be removed; so that every such equation can be reduced to one of the following forms:

$$z^3 = -pz+q \qquad z^3 = +pz+q \qquad z^3 = +pz-q$$

Now, if we have $z^3 = -pz+q$, the rule, attributed by Cardan[236] to one Scipio Ferreus, gives us the root

$$\sqrt[3]{\frac{1}{2}q + \sqrt{\frac{1}{4}q^2 + \frac{1}{27}p^3}} - \sqrt[3]{-\frac{1}{2}q + \sqrt{\frac{1}{4}q^2 + \frac{1}{27}p^3}}. \quad [237]$$

Similarly, when we have $z^3 = +pz+q$ where the square of half the last term is greater than the cube of one-third the coefficient of the next to the last term, the corresponding rule gives us the root

$$\sqrt[3]{\frac{1}{2}q + \sqrt{\frac{1}{4}q^2 - \frac{1}{27}p^3}} + \sqrt[3]{\frac{1}{2}q - \sqrt{\frac{1}{4}q^2 - \frac{1}{27}p^3}}.$$

It is now clear that all problems of which the equations can be reduced to either of these two forms can be constructed without the use of the conic sections except to extract the cube roots of certain known quantities, which process is equivalent to finding two mean proportionals between such a quantity and unity. Again, if we have $z^3 = +pz+q$, where the square of half the last term is not greater than the cube of one-third the coefficient of the next to the last term, describe the circle NQPV with radius NO equal to $\sqrt{\frac{1}{3}p}$, that is to the mean proportional between unity and one-third the known quantity p. Then take $NP = \frac{3q}{p}$, that is, such that NP is to q, the other known

[236] Cardan; Liber X, Cap. XI, fol. 29: "Scipio Ferreus Bononiensis iam annis ab hinc triginta fermè capitulum hoc inuenit, tradidit uero Anthonio Mariæ Florido Veneto, qui cũ in certamen cũ Nicolao Tartalea Brixellense aliquando uenisset, occasionem dedit, ut Nocolaus inuenerit & ipse, qui cum nobis rogantibus tradidisser, suppressa demonstratione, freti hoc auxilio, demonstrationem quæliuimus, eamque in modos, quod difficillimum fuit, redactam sic subjecimus."
See also Cantor, Vol. II (1), p. 444; Smith, Vol. II, p. 462.
[237] Descartes wrote this:

$$\sqrt{C. + \frac{1}{2}q + \sqrt{\frac{1}{4}qq + \frac{1}{27}p^3}} + \sqrt{C. \frac{1}{2}q + \sqrt{\frac{1}{4}qq + \frac{1}{27}p^3}}.$$

quantity, as 1 is to $\frac{1}{3}p$, and inscribe NP in the circle. Divide each of the two arcs NQP and NVP into three equal parts, and the required root is the sum of NQ, the chord subtending one-third the first arc, and NV, the chord subtending one-third of the second arc.[238]

Finally, suppose that we have $z^3 = pz - q$. Construct the circle NQPV whose radius NO is equal to $\sqrt{\frac{1}{3}p}$, and let NP, equal to $\frac{3q}{p}$, be inscribed in this circle; then NQ, the chord of one-third the arc NQP, will be the first of the required roots, and NV, the chord of one-third the other arc, will be the second.

An exception must be made in the case in which the square of half the last term is greater than the cube of one-third the coefficient of the next to the last term;[239] for then the line NP cannot be inscribed in the circle, since it is longer than the diameter. In this case, the two

---

[238] It may be noted that the equation $z^3 = 3z - q$ may be obtained from the equation $z^3 = 3z + q$ by transforming the latter into an equation whose roots have the opposite signs. Then the true roots of $z^3 = 3z - q$ are the false roots of $z^3 = 3z + q$ and vice-versa. Therefore FL = NQ + NP is now the true root.

[239] The so-called irreducible case.

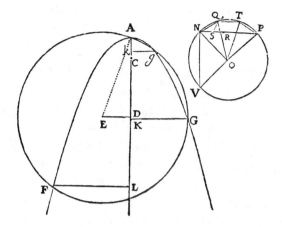

c'eſt a dire qui ſoit à l'autre quantité donnée *q* comme
l'vnité eſt au tiers de *p*; il ne faut que diuiſer chaſcun des
deux arcs N Q P & N V P en trois parties eſgales, & on
aura N Q, la ſubtendue du tiers de l'vn, & N V la ſubtendue du tiers de l'autre, qui iointes enſemble compo-
ſeront la racine cherchée.

Enfin ſi on a $z^3 \infty *p\, z -- q$, en ſuppoſant derechef le
cercle N Q P V, dont le rayon N O ſoit $\sqrt{\tfrac{1}{3}p}$, & l'inſcrite N P ſoit $\frac{3p}{q}$, N Q la ſubtenduë du tiers de l'arc N Q P ſera l'vne des racines cherchées, & N V la ſubtenduë du
tiers de l'autre arc ſera l'autre. Au moins ſi le quarré de
la moitié du dernier terme, n'eſt point plus grand, que le
cube du tiers de la quantité connuë du penultieſme. car
s'il eſtoit plus grand, la ligne N P ne pourroit eſtre inſcrite dans le cercle , a cauſe quelle ſeroit plus longue que
ſon diametre: Ce qui ſeroit cauſe que les deux vrayes racines

cines de cete Equation ne feroient qu'imaginaires , & qu'il ny en auroit de reelles que la fauſſe, qui ſuiuant la reigle de Cardan feroit,

$$\sqrt{C. \tfrac{1}{2}q + \sqrt{\tfrac{1}{4}qq - \tfrac{1}{27}p^3}} + \sqrt{C.\tfrac{1}{2}q - \sqrt{\tfrac{1}{4}qq - \tfrac{1}{27}p^3}}.$$

La façon d'exprimer la valeur de toutes les racines des Equations cubiques: & enſuite de toutes celles qui ne montent que iuſquesau quarré de quarré.

Au reſte il eſt a remarquer que cete façon d'exprimer la valeur des racines par le rapport qu'elles ont aux coſtés de certains cubes dont il n'y a'que le contenu qu'on connoiſſe, n'eſt en rien plus intelligible, ny plus ſimple, que de les exprimer par le rapport qu'elles ont aux ſubtenduës de certains arcs, ou portions de cercles , dont le triple eſt donné. En ſorte que toutes celles des Equations cubiques qui ne peuuent eſtre exprimées par les reigles de Cardan, le peuuent eſtre autant ou plus claireiment par la façon icy propoſée.

Car ſi par exemple , on penſe connoiſtre la racine de cete Equation, $z^3 \infty * - qz + p$. a cauſe qu'on ſçait qu'elle eſt compoſée de deux lignes. dont l'vne eſt le coſté d'vn cube, duquel le contenu eſt $\tfrac{1}{2}q$, adiouſté au coſté d'vn quarré , duquel derechef le contenu eſt $\tfrac{1}{4}qq - \tfrac{1}{27}p^3$; Et l'autre eſt le coſté d'vn autre cube, dont le contenu eſt la difference , qui eſt entre $\tfrac{1}{2}q$, & le coſté de ce quarré dont le contenu eſt $\tfrac{1}{4}qq - \tfrac{1}{27}p^3$, qui eſt tout ce qu'on en apprent par la reigle de Cardan. Il ny a point de doute qu'on ne connoiſſe autant ou plus diſtinctement la racine de celle cy, $z^3 \infty * + q - p$, en la conſiderant inſcrite dans vn cercle, dont le demidiametre eſt $\sqrt{\tfrac{1}{3}p}$, & ſçachant qu'elle y eſt la ſubtenduë d'vn arc dont le triple a pour ſa ſubtendue $\tfrac{3q}{p}$. Meſme ces termes

mes

roots that were true are merely imaginary, and the only real root is the one previously false, which according to Cardan's rule is

$$\sqrt[3]{\frac{1}{2}\,q+\sqrt{\frac{1}{4}\,q^2-\frac{1}{27}\,p^3}} + \sqrt[3]{\frac{1}{2}\,q-\sqrt{\frac{1}{4}\,q^2-\frac{1}{27}\,p^3}}.$$

Furthermore it should be remarked that this method of expressing the roots by means of the relations which they bear to the sides of certain cubes whose contents only are known[240] is in no respect clearer or simpler than the method of expressing them by means of the relations which they bear to the chords of certain arcs (or portions of circles), when arcs three times as long are known. And the roots of the cubic equations which cannot be solved by Cardan's method can be expressed as clearly as any others, or more clearly than the others, by the method given here.

For example, grant that we may consider a root of the equation $z^3 = -qz+p$ known, because we know that it is the sum of two lines of which one is the side of a cube whose volume is $\frac{1}{2}\,q$ increased by the side of a square whose area is $\frac{1}{4}\,q^2-\frac{1}{27}\,p^3$, and the other is the side of another cube whose volume is the difference between $\frac{1}{2}\,q$ and the side of a square whose area is $\frac{1}{4}\,q^2-\frac{1}{27}\,p^3$. This is as much knowledge of the roots as is furnished by Cardan's method. There is no doubt that the value of the root of the equation $z^3 = +qz-p$ is quite as well known and as clearly conceived when it is considered as the length of a chord inscribed in a circle of radius $\sqrt{\frac{1}{3}\,p}$ and subtending an arc that is one-third the arc subtended by a chord of length $\frac{3q}{p}$.

[240] Descartes here makes use of the geometrical conception of finding the cube root of a given quantity.

Indeed, these terms are much less complicated than the others, and they might be made even more concise by the use of some particular symbol to express such chords,[241] just as the symbol $\sqrt[3]{}$ [242] is used to represent the side of a cube.

By methods similar to those already explained, we can express the roots of any biquadratic equation, and there seems to me nothing further to be desired in the matter; for by their very nature these roots cannot be expressed in simpler terms, nor can they be determined by any constuction that is at the same time easier and more general.

It is true that I have not yet stated my grounds for daring to declare a thing possible or impossible, but if it is remembered that in the method I use all problems which present themselves to geometers reduce to a single type, namely, to the question of finding the values of the roots of an equation, it will be clear that a list can be made of all the ways of finding the roots, and that it will then be easy to prove our method the simplest and most general. Solid problems in particular cannot, as I have already said, be constructed without the use of a curve more complex than the circle. This follows at once from the fact that they all reduce to two constructions, namely, to one in which two mean pro-

[241] This is another indication of the tendency of Descartes's age toward symbolism. This suggestion was never adopted.

[242] In Descartes's notation, $\sqrt{}$ C.

mes sont beaucoup moins embarassés que les autres , & ils se trouueront beaucoup plus cours si on veut vser de quelque chiffre particulier pour exprimer ces subten-dües, ainsi qu'on fait du chiffre √C. pour exprimer le costé des cubes.

Et on peut aussy en suite de cecy exprimer les racines de toutes les Equations qui montent iusques au quarré de quarré, par les reigles cy dessus expliquées. En sorte que ie ne sçache rien de plus a desirer en cete matiere. Car enfin la nature de ces racines ne permet pas qu'on les exprime en termes plus simples, ny qu'on les deter-mine par aucune constructiõ qui soit ensemble plus ge-nerale & plus facile.

Il est vray que ie n'ay pas encore dit sur quelles raisons ie me fonde, pour oser ainsi assurer, si vne chose est possi-ble, ou ne l'est pas. Mais si on prent garde comment, par la methode dont ie me sers, tout ce qui tombe sous la consideration des Geometres, se reduist a vn mesme genre de Problesmes, qui est de chercher la valeur des racines de quelque Equation ; on iugera bien qu'il n'est pas malaysé de faire vn dénombrement de toutes les vo-yes par lesquelles on les peut trouuer, qui soit suffisant pour demonstrer qu'on a choisi la plus generale, & la plus simple. Et particulierement pour cequi est des Probles-mes solides, que iay dit ne pouuoir estre construis, sans qu'on y employe quelque ligne plus composée que la circulaire, c'est chose qu'on peut assés trouuer, de ce qu'ils se reduisent tous a deux constructions ; en l'vne desquelles il faut auoir tout ensemble les deux poins,qui determinent deux moyenes proportionelles entre deux

E e e   lignes

lignes données, & en l'autre les deux poins, qui diuiſent
en trois parties eſgales vn arc donné: Car d'autant que la
courbure du cercle ne depend, que d'vn ſimple rapport
de toutes ſes parties, au point qui en eſt le centre ; on ne
peut auſſy s'en ſeruir qu a determiner vn ſeul point entre
deux extremes, comme a trouuer vne moyenne proportionelle entre deux lignes droites données, ou diuiſer en
deux vn arc donné: Au lieu que la courbure des ſections
coniques, dependant touſiours de deux diuerſes choſes,
peut auſſy ſeruir a determiner deux poins differens.

Mais pour cete meſme raiſon il eſt impoſſible, qu'aucun des Probleſmes qui ſont d'vn degré plus compoſés
que les ſolides, & qui preſuppoſent l'inuention de quatre
moyennes proportionelles, ou la diuiſion d'vn angle en
cinq parties eſgales, puiſſent eſtre conſtruits par aucune
des ſections coniques. C'eſt pourquoy ie croyray faire en
cecy tout le mieux qui ſe puiſſe, ſi ie donne vne reigle generale pour les conſtruire, en y employant la ligne courbe qui ſe deſcrit par l'interſectiõ d'vne Parabole & d'vne
ligne droite en la façon cy deſſus expliquée. car i'oſe af-
ſurer qu'il ny en a point de plus ſimple en la nature, qui
puiſſe ſeruir a ce meſme effect ; & vous auéꝫ vû comme
elle ſuit immediatement les ſections coniques, en cete
queſtion tant cherchée par les anciens, dont la ſolution
enſeigne par ordre toutes les lignes courbes, qui doiuent
eſtre receuës en Geometrie.

Vous ſçaues deſia comment, lorſqu'on cherche les
quantités qui ſont requiſes pour la conſtruction de ces
Probleſmes, on les peut touſiours reduire a quelque E-
quation, qui ne monte que iuſques au quarré de cube, ou

au

Façon ge-
nerale
pour con-
ſtruire
tous les
probleſ-
mes re-
duits a

portionals are to be found between two given lines, and one in which two points are to be found which divide a given arc into three equal parts. Inasmuch as the curvature of a circle depends only upon a simple relation between the center and all points on the circumference, the circle can only be used to determine a single point between two extremes, as, for example, to find one mean proportional between two given lines or to bisect a given arc; while, on the other hand, since the curvature of the conic sections always depends upon two different things,[243] it can be used to determine two different points.

For a similar reason, it is impossible that any problem of degree more complex than the solid, involving the finding of four mean proportionals or the division of an angle into five equal parts, can be constructed by the use of one of the conic sections.

I therefore believe that I shall have accomplished all that is possible when I have given a general rule for constructing problems by means of the curve described by the intersection of a parabola and a straight line, as previously explained;[244] for I am convinced that there is nothing of a simpler nature that will serve this purpose. You have seen, too, that this curve directly follows the conic sections in that question to which the ancients devoted so much attention, and whose solution presents in order all the curves that should be received into geometry.

[243] As, for example, the distance of any point from the two foci. Descartes does not say "all points on the circumference," but "toutes ses parties."
[244] See page 84.

When quantities required for the construction of these problems are to be found, you already know how an equation can always be formed that is of no higher degree than the fifth or sixth. You also know how by increasing the roots of this equation we can make them all true, and at the same time have the coefficient of the third term greater than the square of half that of the second term. Also, if it is not higher than the fifth degree it can always be changed into an equation of the sixth degree in which every term is present.

Now to overcome all these difficulties by means of a single rule, I shall consider all these directions applied and the equation thereby reduced to the form:

$$y^6 - py^5 + qy^4 - ry^3 + sy^2 - ty + u = 0$$

in which q is greater than the square of $\frac{1}{2} p$.

au furfolide. Puis vous fçaués auffy comment, en aug- vneEqua-
mentant la valeur des racines de cete Equation, on peut tion qui n'a point
toufiours faire qu'elles deuienent toutes vrayes; & auec plus de
cela que la quãtité connuë du troifiefme terme foit plus fix di-
grande que le quarré de la moitié de celle du fecond: Et méfions.
enfin comment, fi elle ne monte que iufques au furfolide, on la peut hauffer iufques au quarré de cube; & faire que la place d'aucun de fes termes ne manque d'eftre
remplie. Or affin que toutes les difficultés, dont il eft
icy queftion, puiffent eftre refoluës par vne mefme reigle, ie defire qu'on face toutes ces chofes, & par ce
moyen qu'on les reduife toufiours a vne Equation de
telle forme,

$$y^6 -- p y^5 + q y^4 -- r y^3 + s yy -- t y + v \infty 0,$$

& en laquelle la quantité nommée q foit plus grande
que le quarré de la moitié de celle qui eft nommée p.

Eec 2　　　　　　Puis

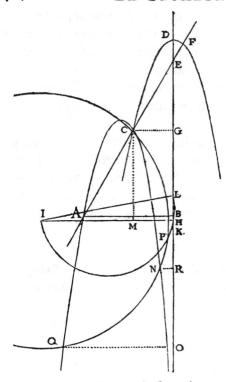

Puis ayant faît a ligne B K indefiniement longue des deux coſtés; & du point B ayant tiré la perpendiculaire A B, dontla longueur ſoit $\frac{1}{2}p$;il faut dans vn plan ſeparé deſcrire vne Parabole, comme C D F dont le coſté droit principalſoit

$$\sqrt{\frac{z}{vv} + q - \frac{1}{4}pp},$$

que ie nommeray n pour abreger. Aprés cela il faut poſer le plan dans lequel eſt cete Parabole ſur celuy ou ſont les lignes A B & B K, en ſorte que ſon aiſſieu D E ſe rencontre iuſtement au deſſus de la ligne droite B K : Et ayant pris la partie de cet aiſſieu, qui eſt entre les poins E & D, eſgale à $\frac{2vv}{pn}$, il faut appliquer ſur ce point E vne longue reigle, en telle façon qu'eſtant auſſy appliquée ſur le point A du plan de deſſous, elle demeure touſiours iointe a ces deux poins, pendant qu'on hauſſera ou baiſſera la Parabole

Produce BK indefinitely in both directions, and at B draw AB perpendicular to BK and equal to $\frac{1}{2} p$. In a separate plane[245] describe the parabola CDF whose principal parameter is

$$\sqrt{\frac{t}{\sqrt{u}} + q - \frac{1}{4} p^2}$$

which we shall represent by n.

Now place the plane containing the parabola on that containing the lines AB and BK, in such a way that the axis DE of the parabola falls along the line BK. Take a point E such that $DE = \dfrac{2\sqrt{u}}{pn}$ and place a ruler so as to connect this point E and the point A of the lower plane. Hold the ruler so that it always connects these points, and slide the parabola up or down, keeping its axis always along BK. Then the

[245] This does not mean in a fixed plane intersecting the first, but, for example, on another piece of paper.

point C, the intersection of the parabola and the ruler, will describe the curve ACN, which is to be used in the construction of the proposed problem.

Having thus described the curve, take a point L in the line BK on the concave side of the parabola, and such that $BL = DE = \dfrac{2\sqrt{u}}{pn}$; then lay off on BK, toward B, LH equal to $\dfrac{t}{2n\sqrt{u}}$, and from H draw HI perpendicular to LH and on the same side as the curve ACN. Take HI equal to

$$\frac{r}{2n^2} + \frac{\sqrt{u}}{n^2} + \frac{pt}{4n^2\sqrt{u}}$$

which we may, for the sake of brevity, set equal to $\dfrac{m}{n^2}$. Join L and I, and describe the circle LPI on LI as diameter; then inscribe in this circle the line LP equal to $\sqrt{\dfrac{s+p\sqrt{u}}{n^2}}$. Finally, describe the circle PCN about I as center and passing through P. This circle will cut or touch the curve ACN in as many points as the equation has roots; and hence the perpendiculars CG, NR, QO, and so on, dropped from these points upon BK, will be the required roots. This rule never fails nor does it admit of any exceptions.

For if the quantity s were so large in proportion to the others, p, q, r, t, u, that the line LP was greater than the diameter of the circle

bole tout le long de la ligne B K , fur laquelle fon aiffieu
eſt appliqué au moyen dequoy l'interſeċtion de cete Parabole, & de cete reigle, qui ſe fera au point C , deſcrira
la ligne courbe A C N, qui eſt celle dont nous auons beſoin de nous ſeruir pour la conſtruċtion du Problesme
propoſé. Car aprés qu'élle eſt ainſi deſcrite, ſi on prent
le point L en la ligne B K, du coſté vers lequel eſt tourné
le ſommet de la Parabole , & qu'on face B L eſgale à D

E, c'eſt à dire à $\frac{\frac{1}{2}Vv}{pn}$: Puis du point L , vers B , qu'on

prene en la meſme ligne B K , la ligne L H, eſgale à

$\frac{t}{2nVv}$; & que du point H ainſi trouué, on tire à angles

droits, du coſté qu'eſt la courbe A C N , la ligne H I,

dont la longeur ſoit $\frac{r}{2nn} + \frac{Vv}{nn} + \frac{pt}{4nnVv}$, qui pour abreger

fera nommée $\frac{m}{nn}$: Et aprés, ayant ioint les poins L & I,
qu'on deſcriue le cercle L P I, dont I L ſoit le diametre;
& qu'on inſcriue en ce cercle la ligne L P dont la lon-

geur ſoit $\frac{\sqrt{s \mp p\,Vv}}{nn}$: Puis enfin du centre I, par le point P

ainſi trouué, qu'on déſcriue le cercle P C N. Ce cercle
couppera ou touchera la ligne courbe A C N , en autant
de poins qu'il y aura de racines en l'Equation : En ſorte
que les perpendiculaires tirées de ces poins ſur la ligne
B K, comme C G, N R, Q O, & ſemblables , ſeront les
racines cherchées. Sans qu'il y ait aucune exception ny
aucun deffaut en cete reigle. Car ſi la quantité s eſtoit
ſi grande, à proportion des autres p, q, r, t, & v, que la ligne L P ſe trouuaſt plus grande que le diametre du cer-

E e e 3

cle

cle I L, en forte qu'elle n'y puft eftre infcrite, il ny auroit
aucune racine en l'Equation propofée qui ne fuft imagi-
naire: Non plus que fi le cercle I P eftoit fi petit, qu'il ne
coupaft la courbe A C N en aucun point.   Et il la peut
couper en fix differens , ainfi qu'il peut y auoir fix
diuerfes racines en l'Equation.   Mais lorfqu'il la coupe
en moins , cela tefmoigne qu'il y a quelques vnes de
ces racines qui font efgales entre elles , oubien qui ne
font qu'imaginaires.

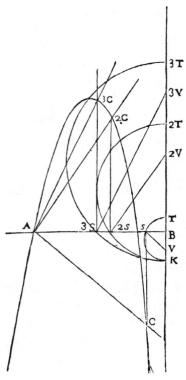

Que

LI,[246] so that LP could not be inscribed in it, every root of the proposed equation would be imaginary; and the same would be true if the circle IP[247] were so small that it did not cut the curve ACN at any point. The circle IP will in general cut the curve ACN in six different points, so that the equation can have six distinct roots.[248] But if it cuts it in fewer points, this indicates that some of the roots are equal or else imaginary.

[246]That is, the circle IPL, of which the diameter is t, page 222.

[247] That is, the circle PCN.

[248] The points determining these roots must be points of intersection of the circle with the main branch of the curve obtained, that is, of the branch ACN.

If, however, this method of tracing the curve ACN by the translation of a parabola seems to you awkward, there are many other ways of describing it. We might take AB and BL as before (page 226), and BK equal to the latus rectum of the parabola, and describe the semicircle KST with its center in BK and cutting AB in some point S. Then from the point T where it ends, take TV toward K equal to BL and join S and V. Draw AC through A parallel to SV, and draw SC through S parallel to BK; then C, the intersection of AC and SC will be one point of the required curve. In this way we can find as many points of the curve as may be desired.

Que si la'façon de tracer la ligne A C N par le mouuement d'vne Parabole vous semble incommode, il est aysé de trouuer plusieurs autres moyens pour la descrire.
Comme si ayant les mesmes quantités que deuant pour
A B & B L; & la mesme pour B K, qu'on auoit posée pour
le costé droit principal de la Parabole; on descrit le demi-
cercle K S T dont le centre soit pris a discretion dans la
ligne B K, en sorte qu'il couppe quelq; part la ligne A B,
comme au point S, & que du point T, ou il finist, on prene vers K la ligne T V, esgale à B L; puis ayant tiré la ligne S V, qu'on en tire vne autre, qui luy soit parallele,
par le point A, comme A C; & qu'on en tire aussy vne
autre par S, qui soit parallele a B K, comme S C; le point
C, ou ces deux paralleles se rencontrent, sera l'vn de ceux
de la ligne courbe cherchée. Et on en peut trouuer, en
mesme sorte, autant d'autres qu'on en desire.

Or

Or la demonſtration de tout cecy eſt aſſés facile. car appliquant la reigle A E auec la Parabole E D ſur le point C; comme il eſt certain qu'elles peuuent y eſtre appliquées enſemble , puiſque ce point C eſt en la courbe A C N, qui eſt deſcrite par leur interſection ; ſi C G ſe nomme y, G D ſera $\frac{yy}{n}$, à cauſe que le coſté droit, qui eſt n, eſt à C G, comme C G a G D. & oſtant D E, qui eſt $\frac{2Vv}{pn}$, de G D, on a $\frac{yy}{n} - \frac{2Vv}{pn}$, pour G E. Puis à cauſe que

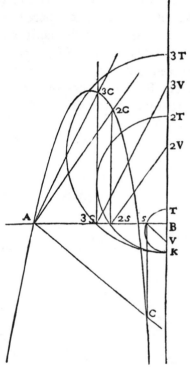

A B eſt a B E, comme C G eſt a G E ; A B eſtant $\frac{1}{2}p$, B E eſt $\frac{py}{2n} - \frac{Vv}{ny}$.

Et tout de meſme en ſuppoſant que le point C de la courbe à eſté trouué par l'interſectiõ des lignes droites, S C parallele à B K, & A C parallele a S V. S B qui eſt eſgale à C G, eſt y : & B K eſtant eſgale au coſté droit de la Parabole, que iay nommé n, B T eſt $\frac{yy}{n}$. car comme K B eſt a B S, ainſi B S eſt a B T. Et T V eſtant

The demonstration of all this is very simple. Place the ruler AE and the parabola FD so that both pass through the point C. This can always be done, since C lies on the curve ACN which is described by the intersection of the parabola and the ruler. If we let $CG = y$, GD will equal $\frac{y^2}{n}$, since the latus rectum n is to CG as CG is to GD. Then $DE = \frac{2\sqrt{u}}{pn}$, and subtracting DE from GD we have $GE = \frac{y^2}{n} - \frac{2\sqrt{u}}{pn}$. Since AB is to BE as CG is to GE, and AB is equal to $\frac{1}{2}p$, therefore $BE = \frac{py}{2n} - \frac{\sqrt{u}}{ny}$. Now let C be a point on the curve generated by the intersection of the line SC, which is parallel to BK, and AC, which is parallel to SV. Let $SB = CG = y$, and $BK = n$, the latus rectum of the parabola. Then $BT = \frac{y^2}{n}$, for KB is to BS as BS is to BT, and since $TV = BL = \dfrac{2\sqrt{u}}{pn}$ we have $BV = \dfrac{y^2}{n} - \dfrac{2\sqrt{u}}{pn}$. Also SB

is to BV as AB is to BE, whence $BE = \dfrac{py}{2n} - \dfrac{\sqrt{u}}{ny}$ as before. It is evident, therefore, that one and the same curve is described by these two methods.

Furthermore, $BL = DE$, and therefore $DL = BE$; also $LH = \dfrac{t}{2n\sqrt{u}}$

and
$$DL = \frac{py}{2n} - \frac{\sqrt{u}}{ny}$$

therefore
$$DH = LH + DL = \frac{py}{2n} - \frac{\sqrt{u}}{ny} + \frac{t}{2n\sqrt{u}}.$$

Also, since $GD = \dfrac{y^2}{n}$,

$$GH = DH - GD = \frac{py}{2n} - \frac{\sqrt{u}}{ny} + \frac{t}{2n\sqrt{u}} - \frac{y^2}{n}$$

which may be written

$$GH = \frac{-y^3 + \dfrac{1}{2}py^2 + \dfrac{ty}{2\sqrt{u}} - \sqrt{u}}{ny}$$

and the square of GH is equal to

$$\frac{y^6 - py^5 + \left(\dfrac{1}{4}p^2 - \dfrac{t}{\sqrt{u}}\right)y^4 + \left(2\sqrt{u} + \dfrac{pt}{2\sqrt{u}}\right)y^3 + \left(\dfrac{t^2}{4u} - p\sqrt{u}\right)y^2 - ty + u}{n^2 y^2}$$

Whatever point of the curve is taken as C, whether toward N or toward Q, it will always be possible to express the square of the segment of BH between the point H and the foot of the perpendicular from C to BH in these same terms connected by these same signs.

eſtant la meſme que BL, c'eſt a dire $\frac{2Vv}{pn}$, BV eſt $\frac{yy}{n} -- \frac{2Vv}{pn}$: & comme SB eſt a BV, ainſi AB eſt à BE, qui eſt par conſequent $\frac{py}{2n} -- \frac{Vv}{ny}$ comme deuant, d'où on voit que c'eſt vne meſme ligne courbe qui ſe deſcrit en ces deux façons.

Aprés cela, pourceque BL & DE ſont eſgales, DL & BE le ſont auſſy: de façon qu'adiouſtãt LH, qui eſt $\frac{t}{2nVv}$, à DL, qui eſt $\frac{py}{2n} -- \frac{Vv}{ny}$, on à la toute DH, qui eſt $\frac{py}{2n} -- \frac{Vv}{ny} + \frac{t}{2nVv}$; & en oſtant GD, qui eſt $\frac{yy}{n}$ on à GH, qui eſt $\frac{py}{2n} -- \frac{Vv}{ny} + \frac{t}{2nVv} -- \frac{yy}{n}$. Ce que i'eſcris par ordre en cete ſorte $GH \infty -- y^3 + \frac{1}{2}pyy + \frac{ty}{2Vv} -- Vv$.

Et le quàrré de GH eſt,

$$y^6 -- py^5 -- \frac{t}{Vv} \Big\} y^4 + 2Vv \Big\} y^3 -- pVv \Big\} yy -- ty + v$$
$$+ \frac{1}{4}pp \qquad + \frac{pt}{2Vv} \qquad + \frac{tt}{4v}$$
$$\overline{\qquad\qquad nn\,yy \qquad\qquad}$$

Et en quelque autre endroit de cete ligne courbe qu'on veuille imaginer le point C, comme vers N, ou vers Q, on trouuera touſiours que le quarré de là ligne droite, qui eſt entre le point H & celuy où tombe la perpendiculaire du point C ſur BH, peut eſtre exprimé en ces meſmes termes, & auec les meſmes ſignes $+$ & $--$.

De plus IH eſtant $\frac{m}{nn}$, & LH eſtant $\frac{t}{2nVv}$, IL eſt $\sqrt{\frac{mm}{n^4} + \frac{tt}{2nVv}}$, à cauſe de l'angle droit IHL; & LP eſtãt $\sqrt{}$

Fff

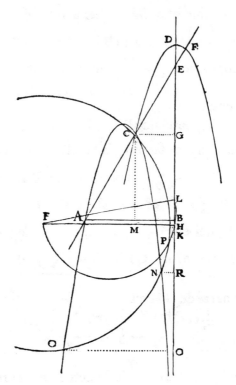

$$\sqrt{\frac{s}{nn} + \frac{pVv}{nn}}, \qquad \text{I P ou I C eft,}$$

$$\sqrt{\frac{mm}{n^4} + \frac{tt}{4.nnv} - \frac{s}{nn} - \frac{pVv}{nn}}, \text{ a caufe auffy de l angle}$$

droit I P L. Puis ayant fait C M perpendiculaire fur I H,
I M eft la difference qui eft entre I H, & H M ou C G,
c'eft a dire entre $\frac{m}{nn}$, & y , en forte que fon quarré
eft toufiours $\frac{mm}{n^4} - \frac{2\,my}{nn} + yy$, qui eftant ofté du quarré

de

Again, $IH = \dfrac{m}{n^2}$, $LH = \dfrac{t}{2n \sqrt{u}}$, whence

$$IL = \sqrt{\dfrac{m^2}{n^4} + \dfrac{t^2}{4n^2 u}},$$

since the angle IHL is a right angle; and since

$$LP = \sqrt{\dfrac{s}{n^2} + \dfrac{p \sqrt{u}}{n^2}}$$

and the angle IPL is a right angle,

$$IC = IP = \sqrt{\dfrac{m^2}{n^4} + \dfrac{t^2}{4n^2 u} - \dfrac{s}{n^2} - \dfrac{p \sqrt{u}}{n^2}}.$$

Now draw CM perpendicular to IH, and

$$IM = HI - HM = HI - CG = \dfrac{m}{n^2} - y;$$

whence the square of IM is $\dfrac{m^2}{n^4} - \dfrac{2my}{n^2} + y^2$.

Taking this from the square of IC there remains the square of CM, or

$$\frac{t^2}{4n^2u} - \frac{s}{n^2} - \frac{p\sqrt{u}}{n^2} + \frac{2my}{n^2} - y^2,$$

and this is equal to the square of GH, previously found. This may be written

$$\frac{-n^2y^4 + 2my^3 - p\sqrt{u}\,y^2 - sy^2 + \frac{t^2}{4u}y^2.}{n^2y^2}$$

Now, putting

$$\frac{t}{\sqrt{u}}y^4 + qy^4 - \frac{1}{4}p^2y^4$$

for n^2y^4, and

$$ry^3 + 2\sqrt{u}\,y^3 + \frac{pt}{2\sqrt{u}}y^3$$

for $2my^3$, and multiplying both members by n^2y^2, we have

$$y^6 - py^5 + \left(\frac{1}{4}p^2 - \frac{t}{\sqrt{u}}\right)y^4 + \left(2\sqrt{u} + \frac{pt}{2\sqrt{u}}\right)y^3 + \left(\frac{t^2}{4u} - p\sqrt{u}\right)y^2 - ty + u$$

equals

$$\left(\frac{1}{4}p^2 - q - \frac{t}{\sqrt{u}}\right)y^4 + \left(r + 2\sqrt{u} + \frac{pt}{2\sqrt{u}}\right)y^3 + \left(\frac{t^2}{4u} - s - p\sqrt{u}\right)y^2,$$

or

$$y^6 - py^5 + qy^4 - ry^3 + sy^2 - ty + u = 0,$$

whence it appears that the lines CG, NR, QO, etc., are the roots of this equation.

If then it be desired to find four mean proportionals between the lines a and b, if we let x be the first, the equation is $x^5 - a^4b = 0$ or $x^6 - a^4bx = 0$. Let $y - a = x$, and we get

$$y^6 - 6ay^5 + 15a^2y^4 - 20a^3y^3 + 15a^4y^2 - (6a^5 + a^4b)y + a^6 + a^5b = 0.$$

Therefore, we must take AB $= 3a$, and BK, the latus rectum of the

de I C, il refte $\frac{tt}{4nnv}$ -- $\frac{s}{nn}$ -- $\frac{p\sqrt{v}}{nn}$ + $\frac{2my}{nn}$ -- yy.

pour le quarré de C M, qui eſt eſgal au quarré de G H deſia trouué. Oubien en faiſant que cete ſomme ſoit diuiſée comme l'autre par $nnyy$, on a

$$-- nny^4 + 2my^3 -- p\sqrt{v}\,yy -- syy + \frac{tt}{4v}yy. \quad \text{Puis}$$
$$nnyy$$

remettant $\frac{t}{\sqrt{v}}y^4 + qy^4 -- \frac{1}{4}ppy^4$, pour nny^4 ; & $ry^3 + 2\sqrt{v}\,y^3 + \frac{pt}{2\sqrt{v}}y^3$, pour $2my^3$: & multipliant l'vne & l'autre ſomme par $nnyy$, on a

$$y^6 -- py^5 -- \left.\begin{matrix}\frac{t}{\sqrt{v}}\\+\frac{1}{4}pp\end{matrix}\right\}y^4 + \left.\begin{matrix}2\sqrt{v}\\+\frac{pt}{2\sqrt{v}}\end{matrix}\right\}y^3 -- \left.\begin{matrix}p\sqrt{v}\\+\frac{tt}{4v}\end{matrix}\right\}yy -- ty + v$$

eſgal à

$$-- \left.\begin{matrix}\frac{t}{\sqrt{v}}\\-- q\\+\frac{1}{4}pp\end{matrix}\right\}y^4 + \left.\begin{matrix}+r\\+2\sqrt{v}\\+\frac{pt}{2\sqrt{v}}\end{matrix}\right\}y^3 -- \left.\begin{matrix}-- p\sqrt{v}\\-- s\\+\frac{tt}{4v}\end{matrix}\right\}yy$$

C'eſt a dire qu'on a,

$$y^6 -- py^5 + qy^4 -- ry^3 + syy -- ty + v \infty 0.$$

D'où il paroiſt que les lignes C G, N R, Q O, & ſemblables ſont les racines de cete Equation, qui eſt ce qu'il falloit demonſtrer.

Ainſi donc ſi on veut trouuer quatre moyennes proportionelles entre les lignes a & b, ayant poſé x pour la premiere , l'Equation eſt $x^5 **** -- a^4b \infty 0$ oubien $x^6 **** -- a + bx^4 \infty 0$. Et faiſant $y -- a \infty x$ il vient

$$y^6 -- 6ay^5 + 15aay^4 -- 20a^3y^3 + 15a^4yy \left.\begin{matrix}-- 6a^5\\-- a^4b\end{matrix}\right\}y \begin{matrix}+a^6\\+a^5b\end{matrix} \infty 0.$$

C'eſt pourquoy il faut prendre $3a$ pour la ligne A B, & $\dfrac{\sqrt{6a^3 + aab}}{\sqrt{aa + ab}} + 6aa$ pour B K, ou le coſté droit de la Pa-

rabole que iay nommé n. & $\frac{2a}{3n}$ $\sqrt{aa+ab}$ pour D E ou
B L. Et aprés auoir defcrit la ligne courbe A C N fur
la mefure de ces trois, il faut faire L H, ∞ $\dfrac{6a^3+aab}{2n\sqrt{aa+ab}}$.

& H I ∞ $\dfrac{10a^3}{nn}+\dfrac{aa}{nn}$ $\sqrt{aa+ab}+\dfrac{18a^4+3a^3b}{nn\sqrt{aa+ab}}$ & L P ∞

$\dfrac{\sqrt{15a^4+6a^3\sqrt{aa+ab}}}{nn}$　Car le cercle qui ayant fon centre
au point I paffera par le point P ainfi trouué, couppera la
courbe aux deux poins C & N ; defquels ayant tiré les
perdendiculaires N R & C G, fi la moindre, N R, eft
oftée de la plus grande, C G, le refte fera, x, la premiere
des quatre moyennes proportionelles cherchées.

Il eft ayfé en mefme façon de diuifer vn angle en cinq
parties efgales, & d'infcrire vne figure d'vnze ou treze
coftés efgaux dans vn cercle, & de trouuer vne infinité
d'autres exemples de cete reigle.

Toutefois il eft a remarquer, qu'en plufieurs de ces
exemples, il peut arriuer que le cercle couppe fi obli-
quement la parabole du fecond genre; que le point de
leur interfection foit difficile a reconnoistre: & ainfi que
cete conftruction ne foit pas commode pour la pratique.
A quoy il feroit ayfé de remedier en compofant d'autres
regles, à l'imitation de celle cy , comme on en peut
compofer de mille fortes.

Mais mon deffein n'eft pas de faire vn gros liure, &
ie tafche plutoft de comprendre beaucoup en peu de
mots: comme on iugera peuteftre que iay fait, fi on con-
fidere, qu'ayant reduit à vne mefme conftruction tous
les

parabola must be

$$\sqrt{\frac{6a^3+a^2b}{\sqrt{a^2+ab}}+6a^2}$$

which I shall call n, and DE or BL will be

$$\frac{2a}{3n}\sqrt{a^2+ab}.$$

Then having described the curve ACN, we must have

$$LH=\frac{6a^3+a^2b}{2n\sqrt{a^2+ab}}$$

and

$$HI=\frac{10a^3}{n^2}+\frac{a^2}{n^2}\sqrt{a^2+ab}+\frac{18a^4+3a^3b}{2n^2\sqrt{a^2+ab}},$$

and

$$LP=\frac{a}{n}\sqrt{15a^2+6a\sqrt{a^2+ab}}.$$

For the circle about I as center will pass through the point P thus found, and cut the curve in the two points C and N. If we draw the perpendiculars NR and CG, and subtract NR, the smaller, from CG, the greater, the remainder will be x, the first of the four required mean proportionals.[249]

This method applies as well to the division of an angle into five equal parts, the inscription of a regular polygon of eleven or thirteen sides in a circle, and an infinity of other problems. It should be remarked, however, that in many of these problems it may happen that the circle cuts the parabola of the second class so obliquely[250] that it is hard to determine the exact point of intersection. In such cases this construction is not of practical value.[251] The difficulty could easily be overcome by forming other rules analogous to these, which might be done in a thousand different ways.

[249] The two roots of the above equation in y are NR and CG. But we know that a is one of the roots of this equation, and therefore NR, the shorter length, must be a, and CG must be y. Then $x = y - a = CG - NR$, the first of the required mean proportionals. Rabuel, p. 580.

[250] That is, makes so small an angle with it.

[251] This is especially noticeable when there are six real positive roots.

But it is not my purpose to write a large book. I am trying rather to include much in a few words, as will perhaps be inferred from what I have done, if it is considered that, while reducing to a single construction all the problems of one class, I have at the same time given a method of transforming them into an infinity of others, and thus of solving each in an infinite number of ways; that, furthermore, having constructed all plane problems by the cutting of a circle by a straight line, and all solid problems by the cutting of a circle by a parabola; and, finally, all that are but one degree more complex by cutting a circle by a curve but one degree higher than the parabola, it is only necessary to follow the same general method to construct all problems, more and more complex, ad infinitum; for in the case of a mathematical progression, whenever the first two or three terms are given, it is easy to find the rest.

I hope that posterity will judge me kindly, not only as to the things which I have explained, but also as to those which I have intentionally omitted so as to leave to others the pleasure of discovery.

[THE END]

les Problefmes d'vn mefme genro, iay tout enfemble donné la façon de les reduire à vne infinité d'autres diuerfes; & ainfi de refoudre chafcun deux en vne infinité de façons. Puis outre cela qu'ayant conftruit tous ceux qui font plans, en coupant d'vn cercle vne ligne droite; & tous ceux qui font folides, en coupant auffy d'vn cercle vne Parabole; & enfin tous ceux qui font d'vn degré plus compofés, en coupant tout de mefme d'vn cercle vne ligne qui n'eft que d'vn degré plus compofée que la Parabole; il ne faut que fuiure la mefme voye pour conftruire tous ceux qui font plus compofés a l'infini. Car en matiere de progreffions Mathematiques, lorfqu'on a les deux ou trois premiers termes, il n'eft pas malayfé de trouuer les autres. Et i'efpere que nos neueux me fçauront gré, non feulement des chofes que iay icy expliquées; mais auffy de celles que iay omifes volontairerement, affin de leur laiffer le plaifir de les inuenter.

F I N.

PAr grace & priuilege du Roy tres chreſtien il eſt permis a l'Autheur du liure intitulé *Diſcours de la Methode &c. plus la Dioptrique, les Meteores, & la Geometrie &c.* de le faire imprimer en telle part que bon luy ſemblera dedans & dehors le royaume de France, & ce pendant le terme de dix annees conſequutiues, a conter du iour qu'il ſera paracheué d'imprimer, ſans qu'aucun autre que le libraire qu'il aura choiſi le puiſſe imprimer, ou faire imprimer, en tout ny en partie, ſous quelque pretexte ou deguiſement que ce puiſſe eſtre; ny en vendre ou debiter d'autre impreſſion que de celle qui aura eſté faite par ſa permiſſion, a peine de mil liures d'amande, confiſcation de tous les exemplaires &c. Ainſi qu'il eſt plus amplement declaré dans les lettres donnees a Paris le 4 iour de May 1637. ſignees par le Roy en ſon conſeil *Ceberet* & ſeellees du grand ſceau de cire iaune ſur ſimple queuë.

l'Autheur a permis a Ian Maire marchand libraire a Leyde, d'imprimer le dit liure & de iouir du dit priuilege pour le tems & aux conditions entre eux accordeés.

Acheué d'imprimer le 8. iour de Iuin 1637.

242

By THE GRACE AND PRIVILEGE of the very Christian King, it is permitted to the author of the book entitled *Discourse on Method,* etc., together with *Dioptrics, Meteorology, and Geometry,* etc., to have printed wherever he wishes, within or without the Kingdom of France, and during the period of ten consecutive years, beginning on the day when the printing is completed, without any publisher (except the one whom he selects) printing it, or causing it to be printed, under any pretext or disguise, or selling or delivering any other impression except that which has been allowed, under penalty of a fine of a thousand livres, the confiscation of all the copies, etc. This is more fully set forth in the letters given at Paris, on the fourth day of May, 1637, signed by the King and his counsel, Ceberet, and sealed with the great seal of yellow wax on a simple ribbon.

The author has given permission to Jan Maire, bookseller at Leyden, to print the said book and enjoy the said privilege for the time and under the conditions agreed upon between them.

The printing is completed the eighth day of June, 1637.

INDEX

A CATALOG OF SELECTED
DOVER BOOKS
IN SCIENCE AND MATHEMATICS

QUALITATIVE THEORY OF DIFFERENTIAL EQUATIONS, V.V. Nemytskii and V.V. Stepanov. Classic graduate-level text by two prominent Soviet mathematicians covers classical differential equations as well as topological dynamics and ergodic theory. Bibliographies. 523pp. 5⅜ × 8½. 65954-2 Pa. $14.95

MATRICES AND LINEAR ALGEBRA, Hans Schneider and George Phillip Barker. Basic textbook covers theory of matrices and its applications to systems of linear equations and related topics such as determinants, eigenvalues and differential equations. Numerous exercises. 432pp. 5⅜ × 8½. 66014-1 Pa. $10.95

QUANTUM THEORY, David Bohm. This advanced undergraduate-level text presents the quantum theory in terms of qualitative and imaginative concepts, followed by specific applications worked out in mathematical detail. Preface. Index. 655pp. 5⅜ × 8½. 65969-0 Pa. $14.95

ATOMIC PHYSICS (8th edition), Max Born. Nobel laureate's lucid treatment of kinetic theory of gases, elementary particles, nuclear atom, wave-corpuscles, atomic structure and spectral lines, much more. Over 40 appendices, bibliography. 495pp. 5⅜ × 8½. 65984-4 Pa. $12.95

ELECTRONIC STRUCTURE AND THE PROPERTIES OF SOLIDS: The Physics of the Chemical Bond, Walter A. Harrison. Innovative text offers basic understanding of the electronic structure of covalent and ionic solids, simple metals, transition metals and their compounds. Problems. 1980 edition. 582pp. 6⅛ × 9¼. 66021-4 Pa. $16.95

BOUNDARY VALUE PROBLEMS OF HEAT CONDUCTION, M. Necati Özisik. Systematic, comprehensive treatment of modern mathematical methods of solving problems in heat conduction and diffusion. Numerous examples and problems. Selected references. Appendices. 505pp. 5⅜ × 8½. 65990-9 Pa. $12.95

A SHORT HISTORY OF CHEMISTRY (3rd edition), J.R. Partington. Classic exposition explores origins of chemistry, alchemy, early medical chemistry, nature of atmosphere, theory of valency, laws and structure of atomic theory, much more. 428pp. 5⅜ × 8½. (Available in U.S. only) 65977-1 Pa. $11.95

A HISTORY OF ASTRONOMY, A. Pannekoek. Well-balanced, carefully reasoned study covers such topics as Ptolemaic theory, work of Copernicus, Kepler, Newton, Eddington's work on stars, much more. Illustrated. References. 521pp. 5⅜ × 8½. 65994-1 Pa. $12.95

PRINCIPLES OF METEOROLOGICAL ANALYSIS, Walter J. Saucier. Highly respected, abundantly illustrated classic reviews atmospheric variables, hydrostatics, static stability, various analyses (scalar, cross-section, isobaric, isentropic, more). For intermediate meteorology students. 454pp. 6⅛ × 9¼. 65979-8 Pa. $14.95

RELATIVITY, THERMODYNAMICS AND COSMOLOGY, Richard C. Tolman. Landmark study extends thermodynamics to special, general relativity; also applications of relativistic mechanics, thermodynamics to cosmological models. 501pp. 5⅜ × 8½. 65383-8 Pa. $13.95

APPLIED ANALYSIS, Cornelius Lanczos. Classic work on analysis and design of finite processes for approximating solution of analytical problems. Algebraic equations, matrices, harmonic analysis, quadrature methods, much more. 559pp. 5⅜ × 8½. 65656-X Pa. $13.95

INTRODUCTION TO ANALYSIS, Maxwell Rosenlicht. Unusually clear, accessible coverage of set theory, real number system, metric spaces, continuous functions, Riemann integration, multiple integrals, more. Wide range of problems. Undergraduate level. Bibliography. 254pp. 5⅜ × 8½. 65038-3 Pa. $8.95

INTRODUCTION TO QUANTUM MECHANICS With Applications to Chemistry, Linus Pauling & E. Bright Wilson, Jr. Classic undergraduate text by Nobel Prize winner applies quantum mechanics to chemical and physical problems. Numerous tables and figures enhance the text. Chapter bibliographies. Appendices. Index. 468pp. 5⅜ × 8½. 64871-0 Pa. $12.95

ASYMPTOTIC EXPANSIONS OF INTEGRALS, Norman Bleistein & Richard A. Handelsman. Best introduction to important field with applications in a variety of scientific disciplines. New preface. Problems. Diagrams. Tables. Bibliography. Index. 448pp. 5⅜ × 8½. 65082-0 Pa. $12.95

MATHEMATICS APPLIED TO CONTINUUM MECHANICS, Lee A. Segel. Analyzes models of fluid flow and solid deformation. For upper-level math, science and engineering students. 608pp. 5⅜ × 8½. 65369-2 Pa. $14.95

ELEMENTS OF REAL ANALYSIS, David A. Sprecher. Classic text covers fundamental concepts, real number system, point sets, functions of a real variable, Fourier series, much more. Over 500 exercises. 352pp. 5⅜ × 8½. 65385-4 Pa. $11.95

PHYSICAL PRINCIPLES OF THE QUANTUM THEORY, Werner Heisenberg. Nóbel Laureate discusses quantum theory, uncertainty, wave mechanics, work of Dirac, Schroedinger, Compton, Wilson, Einstein, etc. 184pp. 5⅜ × 8½. 60113-7 Pa. $6.95

INTRODUCTORY REAL ANALYSIS, A.N. Kolmogorov, S.V. Fomin. Translated by Richard A. Silverman. Self-contained, evenly paced introduction to real and functional analysis. Some 350 problems. 403pp. 5⅜ × 8½. 61226-0 Pa. $10.95

PROBLEMS AND SOLUTIONS IN QUANTUM CHEMISTRY AND PHYSICS, Charles S. Johnson, Jr. and Lee G. Pedersen. Unusually varied problems, detailed solutions in coverage of quantum mechanics, wave mechanics, angular momentum, molecular spectroscopy, scattering theory, more. 280 problems plus 139 supplementary exercises. 430pp. 6½ × 9¼. 65236-X Pa. $13.95

ASYMPTOTIC METHODS IN ANALYSIS, N.G. de Bruijn. An inexpensive, comprehensive guide to asymptotic methods—the pioneering work that teaches by explaining worked examples in detail. Index. 224pp. 5⅜ × 8½.   64221-6 Pa. $7.95

OPTICAL RESONANCE AND TWO-LEVEL ATOMS, L. Allen and J.H. Eberly. Clear, comprehensive introduction to basic principles behind all quantum optical resonance phenomena. 53 illustrations. Preface. Index. 256pp. 5⅜ × 8½.
65533-4 Pa. $8.95

COMPLEX VARIABLES, Francis J. Flanigan. Unusual approach, delaying complex algebra till harmonic functions have been analyzed from real variable viewpoint. Includes problems with answers. 364pp. 5⅜ × 8½.   . 61388-7 Pa. $9.95

ATOMIC SPECTRA AND ATOMIC STRUCTURE, Gerhard Herzberg. One of best introductions; especially for specialist in other fields. Treatment is physical rather than mathematical. 80 illustrations. 257pp. 5⅜ × 8½.   60115-3 Pa. $6.95

APPLIED COMPLEX VARIABLES, John W. Dettman. Step-by-step coverage of fundamentals of analytic function theory—plus lucid exposition of five important applications: Potential Theory; Ordinary Differential Equations; Fourier Transforms; Laplace Transforms; Asymptotic Expansions. 66 figures. Exercises at chapter ends. 512pp. 5⅜ × 8½.   64670-X Pa. $12.95

ULTRASONIC ABSORPTION: An Introduction to the Theory of Sound Absorption and Dispersion in Gases, Liquids and Solids, A.B. Bhatia. Standard reference in the field provides a clear, systematically organized introductory review of fundamental concepts for advanced graduate students, research workers. Numerous diagrams. Bibliography. 440pp. 5⅜ × 8½.   64917-2 Pa. $11.95

UNBOUNDED LINEAR OPERATORS: Theory and Applications, Seymour Goldberg. Classic presents systematic treatment of the theory of unbounded linear operators in normed linear spaces with applications to differential equations. Bibliography. 199pp. 5⅜ × 8½.   64830-3 Pa. $7.95

LIGHT SCATTERING BY SMALL PARTICLES, H.C. van de Hulst. Comprehensive treatment including full range of useful approximation methods for researchers in chemistry, meteorology and astronomy. 44 illustrations. 470pp. 5⅜ × 8½.   64228-3 Pa. $11.95

CONFORMAL MAPPING ON RIEMANN SURFACES, Harvey Cohn. Lucid, insightful book presents ideal coverage of subject. 334 exercises make book perfect for self-study. 55 figures. 352pp. 5⅜ × 8¼.   64025-6 Pa. $11.95

OPTICKS, Sir Isaac Newton. Newton's own experiments with spectroscopy, colors, lenses, reflection, refraction, etc., in language the layman can follow. Foreword by Albert Einstein. 532pp. 5⅜ × 8½.   60205-2 Pa. $11.95

GENERALIZED INTEGRAL TRANSFORMATIONS, A.H. Zemanian. Graduate-level study of recent generalizations of the Laplace, Mellin, Hankel, K. Weierstrass, convolution and other simple transformations. Bibliography. 320pp. 5⅜ × 8½.   65375-7 Pa. $8.95

THE ELECTROMAGNETIC FIELD, Albert Shadowitz. Comprehensive undergraduate text covers basics of electric and magnetic fields, builds up to electromagnetic theory. Also related topics, including relativity. Over 900 problems. 768pp. 5⅜ × 8¼. 65660-8 Pa. $18.95

FOURIER SERIES, Georgi P. Tolstov. Translated by Richard A. Silverman. A valuable addition to the literature on the subject, moving clearly from subject to subject and theorem to theorem. 107 problems, answers. 336pp. 5⅜ × 8½. 63317-9 Pa. $9.95

THEORY OF ELECTROMAGNETIC WAVE PROPAGATION, Charles Herach Papas. Graduate-level study discusses the Maxwell field equations, radiation from wire antennas, the Doppler effect and more. xiii + 244pp. 5⅜ × 8½. 65678-0 Pa. $6.95

DISTRIBUTION THEORY AND TRANSFORM ANALYSIS: An Introduction to Generalized Functions, with Applications, A.H. Zemanian. Provides basics of distribution theory, describes generalized Fourier and Laplace transformations. Numerous problems. 384pp. 5⅜ × 8½. 65479-6 Pa. $11.95

THE PHYSICS OF WAVES, William C. Elmore and Mark A. Heald. Unique overview of classical wave theory. Acoustics, optics, electromagnetic radiation, more. Ideal as classroom text or for self-study. Problems. 477pp. 5⅜ × 8½. 64926-1 Pa. $12.95

CALCULUS OF VARIATIONS WITH APPLICATIONS, George M. Ewing. Applications-oriented introduction to variational theory develops insight and promotes understanding of specialized books, research papers. Suitable for advanced undergraduate/graduate students as primary, supplementary text. 352pp. 5⅜ × 8½. 64856-7 Pa. $9.95

A TREATISE ON ELECTRICITY AND MAGNETISM, James Clerk Maxwell. Important foundation work of modern physics. Brings to final form Maxwell's theory of electromagnetism and rigorously derives his general equations of field theory. 1,084pp. 5⅜ × 8½. 60636-8, 60637-6 Pa., Two-vol. set $23.90

AN INTRODUCTION TO THE CALCULUS OF VARIATIONS, Charles Fox. Graduate-level text covers variations of an integral, isoperimetrical problems, least action, special relativity, approximations, more. References. 279pp. 5⅜ × 8½. 65499-0 Pa. $8.95

HYDRODYNAMIC AND HYDROMAGNETIC STABILITY, S. Chandrasekhar. Lucid examination of the Rayleigh-Benard problem; clear coverage of the theory of instabilities causing convection. 704pp. 5⅜ × 8¼. 64071-X Pa. $14.95

CALCULUS OF VARIATIONS, Robert Weinstock. Basic introduction covering isoperimetric problems, theory of elasticity, quantum mechanics, electrostatics, etc. Exercises throughout. 326pp. 5⅜ × 8½. 63069-2 Pa. $8.95

DYNAMICS OF FLUIDS IN POROUS MEDIA, Jacob Bear. For advanced students of ground water hydrology, soil mechanics and physics, drainage and irrigation engineering and more. 335 illustrations. Exercises, with answers. 784pp. 6⅛ × 9¼. 65675-6 Pa. $19.95

NUMERICAL METHODS FOR SCIENTISTS AND ENGINEERS, Richard Hamming. Classic text stresses frequency approach in coverage of algorithms, polynomial approximation, Fourier approximation, exponential approximation, other topics. Revised and enlarged 2nd edition. 721pp. 5⅜ × 8½.
65241-6 Pa. $15.95

THEORETICAL SOLID STATE PHYSICS, Vol. I: Perfect Lattices in Equilibrium; Vol. II: Non-Equilibrium and Disorder, William Jones and Norman H. March. Monumental reference work covers fundamental theory of equilibrium properties of perfect crystalline solids, non-equilibrium properties, defects and disordered systems. Appendices. Problems. Preface. Diagrams. Index. Bibliography. Total of 1,301pp. 5⅜ × 8½. Two volumes. Vol. I 65015-4 Pa. $16.95
Vol. II 65016-2 Pa. $14.95

OPTIMIZATION THEORY WITH APPLICATIONS, Donald A. Pierre. Broad-spectrum approach to important topic. Classical theory of minima and maxima, calculus of variations, simplex technique and linear programming, more. Many problems, examples. 640pp. 5⅜ × 8½.
65205-X Pa. $14.95

THE CONTINUUM: A Critical Examination of the Foundation of Analysis, Hermann Weyl. Classic of 20th-century foundational research deals with the conceptual problem posed by the continuum. 156pp. 5⅜ × 8½.
67982-9 Pa. $6.95

ESSAYS ON THE THEORY OF NUMBERS, Richard Dedekind. Two classic essays by great German mathematician: on the theory of irrational numbers; and on transfinite numbers and properties of natural numbers. 115pp. 5⅜ × 8½.
21010-3 Pa. $5.95

THE FUNCTIONS OF MATHEMATICAL PHYSICS, Harry Hochstadt. Comprehensive treatment of orthogonal polynomials, hypergeometric functions, Hill's equation, much more. Bibliography. Index. 322pp. 5⅜ × 8½.
65214-9 Pa. $9.95

NUMBER THEORY AND ITS HISTORY, Oystein Ore. Unusually clear, accessible introduction covers counting, properties of numbers, prime numbers, much more. Bibliography. 380pp. 5⅜ × 8½.
65620-9 Pa. $9.95

THE VARIATIONAL PRINCIPLES OF MECHANICS, Cornelius Lanczos. Graduate level coverage of calculus of variations, equations of motion, relativistic mechanics, more. First inexpensive paperbound edition of classic treatise. Index. Bibliography. 418pp. 5⅜ × 8½.
65067-7 Pa. $12.95

MATHEMATICAL TABLES AND FORMULAS, Robert D. Carmichael and Edwin R. Smith. Logarithms, sines, tangents, trig functions, powers, roots, reciprocals, exponential and hyperbolic functions, formulas and theorems. 269pp. 5⅜ × 8½.
60111-0 Pa. $6.95

THEORETICAL PHYSICS, Georg Joos, with Ira M. Freeman. Classic overview covers essential math, mechanics, electromagnetic theory, thermodynamics, quantum mechanics, nuclear physics, other topics. First paperback edition. xxiii + 885pp. 5⅜ × 8½.
65227-0 Pa. $21.95

HANDBOOK OF MATHEMATICAL FUNCTIONS WITH FORMULAS, GRAPHS, AND MATHEMATICAL TABLES, edited by Milton Abramowitz and Irene A. Stegun. Vast compendium: 29 sets of tables, some to as high as 20 places. 1,046pp. 8 × 10½. 61272-4 Pa. $24.95

MATHEMATICAL METHODS IN PHYSICS AND ENGINEERING, John W. Dettman. Algebraically based approach to vectors, mapping, diffraction, other topics in applied math. Also generalized functions, analytic function theory, more. Exercises. 448pp. 5⅜ × 8¼. 65649-7 Pa. $10.95

A SURVEY OF NUMERICAL MATHEMATICS, David M. Young and Robert Todd Gregory. Broad self-contained coverage of computer-oriented numerical algorithms for solving various types of mathematical problems in linear algebra, ordinary and partial, differential equations, much more. Exercises. Total of 1,248pp. 5⅜ × 8½. Two volumes. Vol. I 65691-8 Pa. $14.95
Vol. II 65692-6 Pa. $14.95

TENSOR ANALYSIS FOR PHYSICISTS, J.A. Schouten. Concise exposition of the mathematical basis of tensor analysis, integrated with well-chosen physical examples of the theory. Exercises. Index. Bibliography. 289pp. 5⅜ × 8½. 65582-2 Pa. $8.95

INTRODUCTION TO NUMERICAL ANALYSIS (2nd Edition), F.B. Hildebrand. Classic, fundamental treatment covers computation, approximation, interpolation, numerical differentiation and integration, other topics. 150 new problems. 669pp. 5⅜ × 8½. 65363-3 Pa. $15.95

INVESTIGATIONS ON THE THEORY OF THE BROWNIAN MOVEMENT, Albert Einstein. Five papers (1905–8) investigating dynamics of Brownian motion and evolving elementary theory. Notes by R. Fürth. 122pp. 5⅜ × 8½. 60304-0 Pa. $4.95

CATASTROPHE THEORY FOR SCIENTISTS AND ENGINEERS, Robert Gilmore. Advanced-level treatment describes mathematics of theory grounded in the work of Poincaré, R. Thom, other mathematicians. Also important applications to problems in mathematics, physics, chemistry and engineering. 1981 edition. References. 28 tables. 397 black-and-white illustrations. xvii + 666pp. 6⅛ × 9¼. 67539-4 Pa. $17.95

AN INTRODUCTION TO STATISTICAL THERMODYNAMICS, Terrell L. Hill. Excellent basic text offers wide-ranging coverage of quantum statistical mechanics, systems of interacting molecules, quantum statistics, more. 523pp. 5⅜ × 8½. 65242-4 Pa. $12.95

STATISTICAL PHYSICS, Gregory H. Wannier. Classic text combines thermodynamics, statistical mechanics and kinetic theory in one unified presentation of thermal physics. Problems with solutions. Bibliography. 532pp. 5⅜ × 8½. 65401-X Pa. $12.95

ORDINARY DIFFERENTIAL EQUATIONS, Morris Tenenbaum and Harry Pollard. Exhaustive survey of ordinary differential equations for undergraduates in mathematics, engineering, science. Thorough analysis of theorems. Diagrams. Bibliography. Index. 818pp. 5⅜ × 8½. 64940-7 Pa. $18.95

STATISTICAL MECHANICS: Principles and Applications, Terrell L. Hill. Standard text covers fundamentals of statistical mechanics, applications to fluctuation theory, imperfect gases, distribution functions, more. 448pp. 5⅜ × 8½. 65390-0 Pa. $11.95

ORDINARY DIFFERENTIAL EQUATIONS AND STABILITY THEORY: An Introduction, David A. Sánchez. Brief, modern treatment. Linear equation, stability theory for autonomous and nonautonomous systems, etc. 164pp. 5⅜ × 8¼. 63828-6 Pa. $6.95

THIRTY YEARS THAT SHOOK PHYSICS: The Story of Quantum Theory, George Gamow. Lucid, accessible introduction to influential theory of energy and matter. Careful explanations of Dirac's anti-particles, Bohr's model of the atom, much more. 12 plates. Numerous drawings. 240pp. 5⅜ × 8½. 24895-X Pa. $6.95

THEORY OF MATRICES, Sam Perlis. Outstanding text covering rank, non-singularity and inverses in connection with the development of canonical matrices under the relation of equivalence, and without the intervention of determinants. Includes exercises. 237pp. 5⅜ × 8½. 66810-X Pa. $8.95

GREAT EXPERIMENTS IN PHYSICS: Firsthand Accounts from Galileo to Einstein, edited by Morris H. Shamos. 25 crucial discoveries: Newton's laws of motion, Chadwick's study of the neutron, Hertz on electromagnetic waves, more. Original accounts clearly annotated. 370pp. 5⅜ × 8½. 25346-5 Pa. $10.95

INTRODUCTION TO PARTIAL DIFFERENTIAL EQUATIONS WITH AP-PLICATIONS, E.C. Zachmanoglou and Dale W. Thoe. Essentials of partial differential equations applied to common problems in engineering and the physical sciences. Problems and answers. 416pp. 5⅜ × 8½. 65251-3 Pa. $11.95

BURNHAM'S CELESTIAL HANDBOOK, Robert Burnham, Jr. Thorough guide to the stars beyond our solar system. Exhaustive treatment. Alphabetical by constellation: Andromeda to Cetus in Vol. 1; Chamaeleon to Orion in Vol. 2; and Pavo to Vulpecula in Vol. 3. Hundreds of illustrations. Index in Vol. 3. 2,000pp. 6⅛ × 9¼. 23567-X, 23568-8, 23673-0 Pa., Three-vol. set $44.85

CHEMICAL MAGIC, Leonard A. Ford. Second Edition, Revised by E. Winston Grundmeier. Over 100 unusual stunts demonstrating cold fire, dust explosions, much more. Text explains scientific principles and stresses safety precautions. 128pp. 5⅜ × 8½. 67628-5 Pa. $5.95

AMATEUR ASTRONOMER'S HANDBOOK, J.B. Sidgwick. Timeless, comprehensive coverage of telescopes, mirrors, lenses, mountings, telescope drives, micrometers, spectroscopes, more. 189 illustrations. 576pp. 5⅜ × 8¼. (Available in U.S. only) 24034-7 Pa. $11.95

SPECIAL FUNCTIONS, N.N. Lebedev. Translated by Richard Silverman. Famous Russian work treating more important special functions, with applications to specific problems of physics and engineering. 38 figures. 308pp. 5⅜ × 8½.
60624-4 Pa. $9.95

OBSERVATIONAL ASTRONOMY FOR AMATEURS, J.B. Sidgwick. Mine of useful data for observation of sun, moon, planets, asteroids, aurorae, meteors, comets, variables, binaries, etc. 39 illustrations. 384pp. 5⅜ × 8¼. (Available in U.S. only)
24033-9 Pa. $8.95

INTEGRAL EQUATIONS, F.G. Tricomi. Authoritative, well-written treatment of extremely useful mathematical tool with wide applications. Volterra Equations, Fredholm Equations, much more. Advanced undergraduate to graduate level. Exercises. Bibliography. 238pp. 5⅜ × 8½.
64828-1 Pa. $8.95

POPULAR LECTURES ON MATHEMATICAL LOGIC, Hao Wang. Noted logician's lucid treatment of historical developments, set theory, model theory, recursion theory and constructivism, proof theory, more. 3 appendixes. Bibliography. 1981 edition. ix + 283pp. 5⅜ × 8½.
67632-3 Pa. $8.95

MODERN NONLINEAR EQUATIONS, Thomas L. Saaty. Emphasizes practical solution of problems; covers seven types of equations. ". . . a welcome contribution to the existing literature. . . ."—Math Reviews. 490pp. 5⅜ × 8½. 64232-1 Pa. $11.95

FUNDAMENTALS OF ASTRODYNAMICS, Roger Bate et al. Modern approach developed by U.S. Air Force Academy. Designed as a first course. Problems, exercises. Numerous illustrations. 455pp. 5⅜ × 8½.
60061-0 Pa. $9.95

INTRODUCTION TO LINEAR ALGEBRA AND DIFFERENTIAL EQUATIONS, John W. Dettman. Excellent text covers complex numbers, determinants, orthonormal bases, Laplace transforms, much more. Exercises with solutions. Undergraduate level. 416pp. 5⅜ × 8½.
65191-6 Pa. $10.95

INCOMPRESSIBLE AERODYNAMICS, edited by Bryan Thwaites. Covers theoretical and experimental treatment of the uniform flow of air and viscous fluids past two-dimensional aerofoils and three-dimensional wings; many other topics. 654pp. 5⅜ × 8½.
65465-6 Pa. $16.95

INTRODUCTION TO DIFFERENCE EQUATIONS, Samuel Goldberg. Exceptionally clear exposition of important discipline with applications to sociology, psychology, economics. Many illustrative examples; over 250 problems. 260pp. 5⅜ × 8½.
65084-7 Pa. $8.95

LAMINAR BOUNDARY LAYERS, edited by L. Rosenhead. Engineering classic covers steady boundary layers in two- and three-dimensional flow, unsteady boundary layers, stability, observational techniques, much more. 708pp. 5⅜ × 8½.
65646-2 Pa. $18.95

LECTURES ON CLASSICAL DIFFERENTIAL GEOMETRY, Second Edition, Dirk J. Struik. Excellent brief introduction covers curves, theory of surfaces, fundamental equations, geometry on a surface, conformal mapping, other topics. Problems. 240pp. 5⅜ × 8½.
65609-8 Pa. $8.95

ROTARY-WING AERODYNAMICS, W.Z. Stepniewski. Clear, concise text covers aerodynamic phenomena of the rotor and offers guidelines for helicopter performance evaluation. Originally prepared for NASA. 537 figures. 640pp. 6⅛ × 9¼.
64647-5 Pa. $15.95

DIFFERENTIAL GEOMETRY, Heinrich W. Guggenheimer. Local differential geometry as an application of advanced calculus and linear algebra. Curvature, transformation groups, surfaces, more. Exercises. 62 figures. 378pp. 5⅜ × 8½.
63433-7 Pa. $9.95

INTRODUCTION TO SPACE DYNAMICS, William Tyrrell Thomson. Comprehensive, classic introduction to space-flight engineering for advanced undergraduate and graduate students. Includes vector algebra, kinematics, transformation of coordinates. Bibliography. Index. 352pp. 5⅜ × 8½.   65113-4 Pa. $9.95

A SURVEY OF MINIMAL SURFACES, Robert Osserman. Up-to-date, in-depth discussion of the field for advanced students. Corrected and enlarged edition covers new developments. Includes numerous problems. 192pp. 5⅜ × 8½.
64998-9 Pa. $8.95

ANALYTICAL MECHANICS OF GEARS, Earle Buckingham. Indispensable reference for modern gear manufacture covers conjugate gear-tooth action, gear-tooth profiles of various gears, many other topics. 263 figures. 102 tables. 546pp. 5⅜ × 8½.   65712-4 Pa. $14.95

SET THEORY AND LOGIC, Robert R. Stoll. Lucid introduction to unified theory of mathematical concepts. Set theory and logic seen as tools for conceptual understanding of real number system. 496pp. 5⅜ × 8¼.   63829-4 Pa. $12.95

A HISTORY OF MECHANICS, René Dugas. Monumental study of mechanical principles from antiquity to quantum mechanics. Contributions of ancient Greeks, Galileo, Leonardo, Kepler, Lagrange, many others. 671pp. 5⅜ × 8½.
65632-2 Pa. $14.95

FAMOUS PROBLEMS OF GEOMETRY AND HOW TO SOLVE THEM, Benjamin Bold. Squaring the circle, trisecting the angle, duplicating the cube: learn their history, why they are impossible to solve, then solve them yourself. 128pp. 5⅜ × 8½.   24297-8 Pa. $4.95

MECHANICAL VIBRATIONS, J.P. Den Hartog. Classic textbook offers lucid explanations and illustrative models, applying theories of vibrations to a variety of practical industrial engineering problems. Numerous figures. 233 problems, solutions. Appendix. Index. Preface. 436pp. 5⅜ × 8½.   64785-4 Pa. $11.95

CURVATURE AND HOMOLOGY, Samuel I. Goldberg. Thorough treatment of specialized branch of differential geometry. Covers Riemannian manifolds, topology of differentiable manifolds, compact Lie groups, other topics. Exercises. 315pp. 5⅜ × 8½.   64314-X Pa. $9.95

HISTORY OF STRENGTH OF MATERIALS, Stephen P. Timoshenko. Excellent historical survey of the strength of materials with many references to the theories of elasticity and structure. 245 figures. 452pp. 5⅜ × 8½. 61187-6 Pa. $12.95

GEOMETRY OF COMPLEX NUMBERS, Hans Schwerdtfeger. Illuminating, widely praised book on analytic geometry of circles, the Moebius transformation, and two-dimensional non-Euclidean geometries. 200pp. 5⅜ × 8¼.
63830-8 Pa. $8.95

MECHANICS, J.P. Den Hartog. A classic introductory text or refresher. Hundreds of applications and design problems illuminate fundamentals of trusses, loaded beams and cables, etc. 334 answered problems. 462pp. 5⅜ × 8½. 60754-2 Pa. $10.95

TOPOLOGY, John G. Hocking and Gail S. Young. Superb one-year course in classical topology. Topological spaces and functions, point-set topology, much more. Examples and problems. Bibliography. Index. 384pp. 5⅜ × 8¼.
65676-4 Pa. $10.95

STRENGTH OF MATERIALS, J.P. Den Hartog. Full, clear treatment of basic material (tension, torsion, bending, etc.) plus advanced material on engineering methods, applications. 350 answered problems. 323pp. 5⅜ × 8½. 60755-0 Pa. $9.95

ELEMENTARY CONCEPTS OF TOPOLOGY, Paul Alexandroff. Elegant, intuitive approach to topology from set-theoretic topology to Betti groups; how concepts of topology are useful in math and physics. 25 figures. 57pp. 5⅜ × 8½.
60747-X Pa. $3.95

ADVANCED STRENGTH OF MATERIALS, J.P. Den Hartog. Superbly written advanced text covers torsion, rotating disks, membrane stresses in shells, much more. Many problems and answers. 388pp. 5⅜ × 8½. 65407-9 Pa. $10.95

COMPUTABILITY AND UNSOLVABILITY, Martin Davis. Classic graduate-level introduction to theory of computability, usually referred to as theory of recurrent functions. New preface and appendix. 288pp. 5⅜ × 8½. 61471-9 Pa. $8.95

GENERAL CHEMISTRY, Linus Pauling. Revised 3rd edition of classic first-year text by Nobel laureate. Atomic and molecular structure, quantum mechanics, statistical mechanics, thermodynamics correlated with descriptive chemistry. Problems. 992pp. 5⅜ × 8½. 65622-5 Pa. $19.95

AN INTRODUCTION TO MATRICES, SETS AND GROUPS FOR SCIENCE STUDENTS, G. Stephenson. Concise, readable text introduces sets, groups, and most importantly, matrices to undergraduate students of physics, chemistry, and engineering. Problems. 164pp. 5⅜ × 8½. 65077-4 Pa. $7.95

THE HISTORICAL BACKGROUND OF CHEMISTRY, Henry M. Leicester. Evolution of ideas, not individual biography. Concentrates on formulation of a coherent set of chemical laws. 260pp. 5⅜ × 8½. 61053-5 Pa. $7.95

THE PHILOSOPHY OF MATHEMATICS: An Introductory Essay, Stephan Körner. Surveys the views of Plato, Aristotle, Leibniz & Kant concerning propositions and theories of applied and pure mathematics. Introduction. Two appendices. Index. 198pp. 5⅜ × 8½. 25048-2 Pa. $8.95

THE DEVELOPMENT OF MODERN CHEMISTRY, Aaron J. Ihde. Authoritative history of chemistry from ancient Greek theory to 20th-century innovation. Covers major chemists and their discoveries. 209 illustrations. 14 tables. Bibliographies. Indices. Appendices. 851pp. 5⅜ × 8½. 64235-6 Pa. $18.95

DE RE METALLICA, Georgius Agricola. The famous Hoover translation of greatest treatise on technological chemistry, engineering, geology, mining of early modern times (1556). All 289 original woodcuts. 638pp. 6¾ × 11.
60006-8 Pa. $18.95

SOME THEORY OF SAMPLING, William Edwards Deming. Analysis of the problems, theory and design of sampling techniques for social scientists, industrial managers and others who find statistics increasingly important in their work. 61 tables. 90 figures. xvii + 602pp. 5⅜ × 8½. 64684-X Pa. $15.95

THE VARIOUS AND INGENIOUS MACHINES OF AGOSTINO RAMELLI: A Classic Sixteenth-Century Illustrated Treatise on Technology, Agostino Ramelli. One of the most widely known and copied works on machinery in the 16th century. 194 detailed plates of water pumps, grain mills, cranes, more. 608pp. 9 × 12.
28180-9 Pa. $24.95

LINEAR PROGRAMMING AND ECONOMIC ANALYSIS, Robert Dorfman, Paul A. Samuelson and Robert M. Solow. First comprehensive treatment of linear programming in standard economic analysis. Game theory, modern welfare economics, Leontief input-output, more. 525pp. 5⅜ × 8½. 65491-5 Pa. $14.95

ELEMENTARY DECISION THEORY, Herman Chernoff and Lincoln E. Moses. Clear introduction to statistics and statistical theory covers data processing, probability and random variables, testing hypotheses, much more. Exercises. 364pp. 5⅜ × 8½. 65218-1 Pa. $10.95

THE COMPLEAT STRATEGYST: Being a Primer on the Theory of Games of Strategy, J.D. Williams. Highly entertaining classic describes, with many illustrated examples, how to select best strategies in conflict situations. Prefaces. Appendices. 268pp. 5⅜ × 8½. 25101-2 Pa. $7.95

CONSTRUCTIONS AND COMBINATORIAL PROBLEMS IN DESIGN OF EXPERIMENTS, Damaraju Raghavarao. In-depth reference work examines orthogonal Latin squares, incomplete block designs, tactical configuration, partial geometry, much more. Abundant explanations, examples. 416pp. 5⅜ × 8¼.
65685-3 Pa. $10.95

THE ABSOLUTE DIFFERENTIAL CALCULUS (CALCULUS OF TENSORS), Tullio Levi-Civita. Great 20th-century mathematician's classic work on material necessary for mathematical grasp of theory of relativity. 452pp. 5⅜ × 8½.
63401-9 Pa. $11.95

VECTOR AND TENSOR ANALYSIS WITH APPLICATIONS, A.I. Borisenko and I.E. Tarapov. Concise introduction. Worked-out problems, solutions, exercises. 257pp. 5⅜ × 8¼. 63833-2 Pa. $8.95

THE FOUR-COLOR PROBLEM: Assaults and Conquest, Thomas L. Saaty and Paul G. Kainen. Engrossing, comprehensive account of the century-old combinatorial topological problem, its history and solution. Bibliographies. Index. 110 figures. 228pp. 5⅜ × 8½.   65092-8 Pa. $6.95

CATALYSIS IN CHEMISTRY AND ENZYMOLOGY, William P. Jencks. Exceptionally clear coverage of mechanisms for catalysis, forces in aqueous solution, carbonyl- and acyl-group reactions, practical kinetics, more. 864pp. 5⅜ × 8½.   65460-5 Pa. $19.95

PROBABILITY: An Introduction, Samuel Goldberg. Excellent basic text covers set theory, probability theory for finite sample spaces, binomial theorem, much more. 360 problems. Bibliographies. 322pp. 5⅜ × 8½.   65252-1 Pa. $9.95

LIGHTNING, Martin A. Uman. Revised, updated edition of classic work on the physics of lightning. Phenomena, terminology, measurement, photography, spectroscopy, thunder, more. Reviews recent research. Bibliography. Indices. 320pp. 5⅜ × 8¼.   64575-4 Pa. $8.95

PROBABILITY THEORY: A Concise Course, Y.A. Rozanov. Highly readable, self-contained introduction covers combination of events, dependent events, Bernoulli trials, etc. Translation by Richard Silverman. 148pp. 5⅜ × 8¼.
63544-9 Pa. $6.95

AN INTRODUCTION TO HAMILTONIAN OPTICS, H. A. Buchdahl. Detailed account of the Hamiltonian treatment of aberration theory in geometrical optics. Many classes of optical systems defined in terms of the symmetries they possess. Problems with detailed solutions. 1970 edition. xv + 360pp. 5⅜ × 8½.
67597-1 Pa. $10.95

STATISTICS MANUAL, Edwin L. Crow, et al. Comprehensive, practical collection of classical and modern methods prepared by U.S. Naval Ordnance Test Station. Stress on use. Basics of statistics assumed. 288pp. 5⅜ × 8½.
60599-X Pa. $7.95

DICTIONARY/OUTLINE OF BASIC STATISTICS, John E. Freund and Frank J. Williams. A clear concise dictionary of over 1,000 statistical terms and an outline of statistical formulas covering probability, nonparametric tests, much more. 208pp. 5⅜ × 8½.   66796-0 Pa. $7.95

STATISTICAL METHOD FROM THE VIEWPOINT OF QUALITY CONTROL, Walter A. Shewhart. Important text explains regulation of variables, uses of statistical control to achieve quality control in industry, agriculture, other areas. 192pp. 5⅜ × 8½.   65232-7 Pa. $7.95

THE INTERPRETATION OF GEOLOGICAL PHASE DIAGRAMS, Ernest G. Ehlers. Clear, concise text emphasizes diagrams of systems under fluid or containing pressure; also coverage of complex binary systems, hydrothermal melting, more. 288pp. 6½ × 9¼.   65389-7 Pa. $10.95

STATISTICAL ADJUSTMENT OF DATA, W. Edwards Deming. Introduction to basic concepts of statistics, curve fitting, least squares solution, conditions without parameter, conditions containing parameters. 26 exercises worked out. 271pp. 5⅜ × 8½.   64685-8 Pa. $9.95

TENSOR CALCULUS, J.L. Synge and A. Schild. Widely used introductory text covers spaces and tensors, basic operations in Riemannian space, non-Riemannian spaces, etc. 324pp. 5⅜ × 8¼. 63612-7 Pa. $9.95

A CONCISE HISTORY OF MATHEMATICS, Dirk J. Struik. The best brief history of mathematics. Stresses origins and covers every major figure from ancient Near East to 19th century. 41 illustrations. 195pp. 5⅜ × 8½. 60255-9 Pa. $7.95

A SHORT ACCOUNT OF THE HISTORY OF MATHEMATICS, W.W. Rouse Ball. One of clearest, most authoritative surveys from the Egyptians and Phoenicians through 19th-century figures such as Grassman, Galois, Riemann. Fourth edition. 522pp. 5⅜ × 8½. 20630-0 Pa. $11.95

HISTORY OF MATHEMATICS, David E. Smith. Nontechnical survey from ancient Greece and Orient to late 19th century; evolution of arithmetic, geometry, trigonometry, calculating devices, algebra, the calculus. 362 illustrations. 1,355pp. 5⅜ × 8½. 20429-4, 20430-8 Pa., Two-vol. set $26.90

THE GEOMETRY OF RENÉ DESCARTES, René Descartes. The great work founded analytical geometry. Original French text, Descartes' own diagrams, together with definitive Smith-Latham translation. 244pp. 5⅜ × 8½. 60068-8 Pa. $7.95

THE ORIGINS OF THE INFINITESIMAL CALCULUS, Margaret E. Baron. Only fully detailed and documented account of crucial discipline: origins; development by Galileo, Kepler, Cavalieri; contributions of Newton, Leibniz, more. 304pp. 5⅜ × 8½. (Available in U.S. and Canada only) 65371-4 Pa. $9.95

THE HISTORY OF THE CALCULUS AND ITS CONCEPTUAL DEVELOPMENT, Carl B. Boyer. Origins in antiquity, medieval contributions, work of Newton, Leibniz, rigorous formulation. Treatment is verbal. 346pp. 5⅜ × 8½. 60509-4 Pa. $9.95

THE THIRTEEN BOOKS OF EUCLID'S ELEMENTS, translated with introduction and commentary by Sir Thomas L. Heath. Definitive edition. Textual and linguistic notes, mathematical analysis. 2,500 years of critical commentary. Not abridged. 1,414pp. 5⅜ × 8½. 60088-2, 60089-0, 60090-4 Pa., Three-vol. set $31.85

GAMES AND DECISIONS: Introduction and Critical Survey, R. Duncan Luce and Howard Raiffa. Superb nontechnical introduction to game theory, primarily applied to social sciences. Utility theory, zero-sum games, n-person games, decision-making, much more. Bibliography. 509pp. 5⅜ × 8½. 65943-7 Pa. $12.95

THE HISTORICAL ROOTS OF ELEMENTARY MATHEMATICS, Lucas N.H. Bunt, Phillip S. Jones, and Jack D. Bedient. Fundamental underpinnings of modern arithmetic, algebra, geometry and number systems derived from ancient civilizations. 320pp. 5⅜ × 8½. 25563-8 Pa. $8.95

CALCULUS REFRESHER FOR TECHNICAL PEOPLE, A. Albert Klaf. Covers important aspects of integral and differential calculus via 756 questions. 566 problems, most answered. 431pp. 5⅜ × 8½. 20370-0 Pa. $8.95

CHALLENGING MATHEMATICAL PROBLEMS WITH ELEMENTARY SOLUTIONS, A.M. Yaglom and I.M. Yaglom. Over 170 challenging problems on probability theory, combinatorial analysis, points and lines, topology, convex polygons, many other topics. Solutions. Total of 445pp. 5⅜ × 8½. Two-vol. set.
Vol. I 65536-9 Pa. $7.95
Vol. II 65537-7 Pa. $7.95

FIFTY CHALLENGING PROBLEMS IN PROBABILITY WITH SOLUTIONS, Frederick Mosteller. Remarkable puzzlers, graded in difficulty, illustrate elementary and advanced aspects of probability. Detailed solutions. 88pp. 5⅜ × 8½.
65355-2 Pa. $4.95

EXPERIMENTS IN TOPOLOGY, Stephen Barr. Classic, lively explanation of one of the byways of mathematics. Klein bottles, Moebius strips, projective planes, map coloring, problem of the Koenigsberg bridges, much more, described with clarity and wit. 43 figures. 210pp. 5⅜ × 8½. 25933-1 Pa. $6.95

RELATIVITY IN ILLUSTRATIONS, Jacob T. Schwartz. Clear nontechnical treatment makes relativity more accessible than ever before. Over 60 drawings illustrate concepts more clearly than text alone. Only high school geometry needed. Bibliography. 128pp. 6⅛ × 9¼. 25965-X Pa. $7.95

AN INTRODUCTION TO ORDINARY DIFFERENTIAL EQUATIONS, Earl A. Coddington. A thorough and systematic first course in elementary differential equations for undergraduates in mathematics and science, with many exercises and problems (with answers). Index. 304pp. 5⅜ × 8½. 65942-9 Pa. $8.95

FOURIER SERIES AND ORTHOGONAL FUNCTIONS, Harry F. Davis. An incisive text combining theory and practical example to introduce Fourier series, orthogonal functions and applications of the Fourier method to boundary-value problems. 570 exercises. Answers and notes. 416pp. 5⅜ × 8½. 65973-9 Pa. $11.95

AN INTRODUCTION TO ALGEBRAIC STRUCTURES, Joseph Landin. Superb self-contained text covers "abstract algebra": sets and numbers, theory of groups, theory of rings, much more. Numerous well-chosen examples, exercises. 247pp. 5⅜ × 8½. 65940-2 Pa. $8.95

Prices subject to change without notice.

Available at your book dealer or write for free Mathematics and Science Catalog to Dept. GI, Dover Publications, Inc., 31 East 2nd St., Mineola, N.Y. 11501. Dover publishes more than 175 books each year on science, elementary and advanced mathematics, biology, music, art, literature, history, social sciences and other areas.